HMS *Glory*

The History of A Light Fleet Aircraft Carrier

1942–1961

Peter Barrett

Foreword by
Rear-Admiral B.C.G. Place, VC, CB, CVO, DSC

PARAPRESS LTD
Tunbridge Wells · Kent

In the same INTO BATTLE series:

© Peter Barrett 1996
ISBN 1-898594-36-8

First published in the UK by
PARAPRESS LTD
12 Dene Way
Speldhurst
Tunbridge Wells
Kent TN3 0NX

A catalogue record for this book is available from the British Library

Printed in Great Britain by
Ipswich Book Co. Ltd., Ipswich, Suffolk

Contents

List of Photographs

vi

Foreword

by

Rear Admiral B.C.G. Place, VC, CB, CVO, DSC

HMS *Glory*'s active life covered little more than ten years but it was a period of considerable significance; one might call these years 'Twilight of Empire' when Britain still had world-wide responsibilities but no longer the wealth to support the largest and most powerful navy.

The light fleet carrier was a cost effective compromise between mobility and the exercise of power, and economical operating and running costs – and, as such, HMS *Glory* steamed many thousands of miles in the oceans of the world to meet actual and potential threats.

One might also call this ten years 'The Age of Transition', coming immediately before the leap forward to the most modern systems we recognise today. Radar and communication systems had made great progress since the ship was first designed, but the thermionic valve had not yet been replaced by the micro-chip and printed circuit. The piston engined aeroplane had reached the peak of its performance but had yet to be superseded by the swept wing jet.

And the sailor still slept in his hammock, hemmed in and often overheated by more and more mechanical and electrical equipment. Peter Barrett makes little mention of this, but it is a great tribute to the fortitude of the British sailor that he worked so cheerfully in conditions varying from tropical heat to Arctic cold, to the very peak of professional skills.

Godfrey Place

Introduction

If men could learn from history, what lessons it might teach us. But passion and party blind our eyes, and the light which experience gives is a lantern on the stern, which shines only on the waves behind us!

COLERIDGE 18th DECEMBER 1831

HMS *Glory* was not a handsome ship in the way that battleships, cruisers, or destroyers were. But she was impressive. A floating airbase, with hangars, workshops and repair shops. Her big bare flight deck was 690 feet long. When HMS *Glory* was commissioned in February 1945, naval aviation was increasing in power like a rising wind; a wind that swept all the seas of the world. Naval aircraft had ranged the oceans in search of the enemy; they had attacked German, Japanese and Italian warships. They were the first aircraft in history to sink a warship by dive-bombing, the first to sink a capital ship with torpedoes, and the first to defeat air attacks on a fleet by fighter defence.

The pilots in HMS *Glory* did not have the benefit of angled flight deck or mirror landing aids, and theirs was a difficult and dangerous life. The job of a naval pilot in the 1940s and 1950s reminds one of the story of a once famous pilot of the Royal Flying Corps named Gordon Bell, who suffered from a stammer. After the First World War he became a 'joy-ride' pilot, and a friend once asked him why he never did aerobatics. Solemnly and with difficulty Gordon Bell replied, 'I d-do two s-stunts every t-time I f-fly, t-taking off and l-landing.' A pilot's job in HMS *Glory* was much the same, where the flight deck could pitch and roll in a very disconcerting fashion.

The flight deck seemed so vast and spacious to everyone when they first set foot on it, but appeared frighteningly small when it came to taking off or landing an aircraft. Flying was the hub of life aboard *Glory*, and to achieve and maintain the highest standards possible many hundreds of men were required to keep her complex machinery in efficient running order: seamen, cooks and stewards, stokers, signalmen, wireless operators, writers, aircraft handlers, engine and airframe fitters, armourers, safety equipment and meteorology ratings, dentist, doctor, padre, shipwrights, sailmakers, sick berth attendants, electricians and of course *Glory*'s Royal Marines who manned the anti-aircraft guns and who played in the band – all went to make up the population in the self-contained world that was HMS *Glory*.

Although the ship's career was short, her duties were varied and full of incident and ranged from the repatriation of Prisoners of War, the surrender of Japanese Forces, her record flying achievements during the Korean War, to the ferrying and humanitarian duties towards the end of her service. Aircraft carrier operations during the 1940s and 1950s were always fraught with danger and there were many tragedies, but all those men who served in *Glory* during her commissions, in whatever capacity, will not easily forget the spirit of unity and purpose, the loyalty and good humour, which always seemed to triumph onboard.

1

Utility Aircraft Carrier
(March 1942 – May 1945)

Towards the end of 1941, the Admiralty had decided to adopt the design of a utility aircraft carrier for service with the fleet. The 'COLOSSUS' class of carrier would be much less costly than the previous 'Fleet' carriers, and huge savings would be made by dispensing with armour plated flight and hangar decks. Further reductions in cost were made by confining the ship's aramament to close-range anti-aircraft weapons of 40mm and 20mm calibre only. To increase the speed of building these ships the hulls would be built to Lloyd's mercantile specification up to main deck level with the idea apparently, of converting the class to commercial use after the war.

Another major saving in cost was achieved by having a maximum speed of 25 knots, which meant that only 40,000 hp was required, instead of more than double that figure needed to attain a speed of 30 knots which had been demanded of the earlier 'Fleet' aircraft carriers. Instead of having to design new machinery, the Director of Naval Construction was able to use two sets of light cruiser boilers and turbines of an existing well proved design.

Cost, weight and manpower were saved, and oil fuel capacity reduced by over a thousand tons in comparison with the *Illustrious* for a similar cruising radius.

The result was a carrier with 690ft (usable) flight deck with a 90ft wide landing area. A single hydraulic catapult, capable of launching a 14,000lb aircraft, was fitted on the port side forward. These 'Light Fleets' proved to be extremely successful in service.

The order to proceed with HMS *Glory* was issued by the Admiralty on Wednesday 25 March 1942. Five months later, on Friday 28 August, the ship's keel was laid in the Musgrave Shipyard of Harland & Wolff, Belfast. In the builders' yards ships were usually given numbers to enable them to be identified during construction, as the name was not normally decided until building was almost complete. It was also difficult to paint the ship's name onto each piece of steel, particularly if the name was a long one. HMS *Glory* was job number 12.

There had been three ships in the Royal Navy called *Glory*, (apart from a number of French ships captured in the eighteenth century and which continued to serve in the Royal Navy under their French name *Gloire*). The first *Glory* to see action was a ship of the line of 1,933 tons, carrying 98 guns, which took part in the battle of the '*Glorious 1st of June' 1794*, serving under Admiral Earl Howe. During the battle *Glory* had fifty-two of her crew killed. Next came a fifth rate ship of 1,153 tons, carrying 38 guns, this ship took part in the capture of the islands of Martinique and Guadeloupe, West Indies, in 1809 and 1810 respectively.

The last *Glory* to see action was a battleship of 12,950 tons armed with four 12 inch and twelve 6 inch guns. She took part in the Gallipoli Campaign and the Dardenelles in 1915 in support of the Anzac Division of immortal fame. The log books of the last two ships are available for inspection at the Public Records Office at Kew, although those for 1810 are in poor condition.

During the second world war, the shipyards at Queen's Island, Belfast, were more severely damaged by enemy air attacks than any other British shipyards, yet at no time did production cease in any department, and in the face of almost insuperable difficulties work was pressed forward at an amazing rate. At the time of *Glory*'s launch, three more aircraft carriers, *Warrior, Magnificent*, and *Powerful*, were also under construction and formed an impressive sight in the yards. Fifteen months after the keel was laid, HMS *Glory* was launched by Lady Cynthia Mary Brooke, wife of Northern Ireland's Prime Minister, on Saturday 27 November 1943. The ship slid very smoothly and gracefully into the Musgrave Channel, to the cheering of the thousands of workers who had been responsible for her construction through some very dangerous times.

HMS *Glory* was then towed to her fitting out quay where she spent the whole of 1944. It was the year which saw the arrival of the V-1 'Doodle Bug' flying bombs in June, and then in September, the V-2 rocket bombs. More significant perhaps, was the ending of Japanese seapower with the battle of Leyte Gulf, in October 1944. The following month, on Wednesday 1 November, *Glory*'s first Commanding Officer, Captain Sir Anthony Wass Buzzard, DSO,

OBE, RN, arrived to assume command of the ship. He was tall, ginger haired, and had experienced many sea battles since the war had started. His first command, HMS *Gurkha*, was the first destroyer to be sunk by German air attack on Tuesday 9 April 1940, and Captain Buzzard spent nearly an hour in icy water before being picked up.

Captain Buzzard spent the first three months of his command overseeing the final fitting out of the ship from his office ashore on Thompson's Triangle. Early in February 1945, the new ship's company began to arrive from Devonport, and finally at noon on Wednesday 21 February 1945, HMS *Glory* was officially commissioned. Lower deck was cleared and mustered on the jetty, and the final inspection by the Admiral Superintendent Contract Built Ships (Belfast) was carried out. Hands were piped to dinner at 1230, and at 1310 commissioning cards and mess traps were issued; the rest of the afternoon was spent transporting kit and stores from the Nissen huts on Thompson's Triangle into *Glory*'s spacious hangar. The ship's log for that day shows that no one reported sick!

The next few weeks constituted a shifting kaleidoscope of confusion as all hands were allocated their parts of ship. Official watchkeeping duties commenced, and more stores and ammunition were loaded aboard. Many of the ship's company on that first commission remember the fine hospitality afforded them by the people of Belfast. George Reid, Harry Burrows, and Phil Lister talk with great affection of their time in the city, and Charles Causley, the poet and writer, also a member of the ship's company, wrote several poems during his time in *Glory*, two of which are reproduced in this book. It was an exciting time for the majority who were young Hostilities Only ratings, getting their first taste of life aboard this huge warship, and looking forward perhaps to visiting exotic foreign lands.

The war in Europe was finishing, but there was no end in sight for the war in the Far East. *Glory*'s ship's company were getting anxious, 'buzzes' were rife throughout the ship until Captain Buzzard addressed the ship's company over the tannoy: 'I told you last week that I would give you as much information as I can regarding the future, and I imagine the sort of questions that you have been turning over in your mind have been these: What job are we likely

3

to be required to do? What is the programme likely to be for setting about the job? What will be the effect of that programme on our private affairs? The first question depends largely on how the war goes. It's anybody's guess but those in the best position to guess would, I think say that the German war may be over in a matter of months, but that the Japanese war is unlikely to be over in less than a year.

'That means we may or may not be in time to help finish off Germany, but there is every likelihood of our being ready in time to help with the defeat of Japan. Whether or not we shall in fact go out to the Far East I do not know, and if I did know, I should probably have to say I don't for security reasons. But you probably saw in the papers a few days ago that the Prime Minister has promised that all available forces would be turned onto Japan, as soon as Germany has been beaten, in order to finish off that job as quickly as possible. Now it is fairly generally recognised that aircraft carriers are the things most wanted to beat Japan quickly.

'They are wanted to destroy or suppress her navy, her airforce, her shipping and her army, thus paving the way for the allied armies to land on Japan and all the islands held by the Japanese. The general conclusion is that we may have an important part to play against the Japanese and the quicker we can get down to the job the quicker it will be over. How do we set about it and what's the programme? First of all we have got to learn to live in the ship and make it a comfortable and healthy home. You have already made a good start in that, and I am glad to see that the majority of you are returning from your leave punctually and in good order, I hope you will keep that up. Each individual and each department has got to learn its own job for working the ship at sea. All departments have got to learn to work together as a team in acton, in just the same way as a football team learn to play together.

'There will be several weeks alongside here, then a week or two at sea doing trials, then several weeks working up with the torpedo bomber squadron aboard. After which the fighter bomber squadron is likely to join. In about three months from now we shall probably be expected to be fit to fight. Exactly where we will be during the working up period I don't know but we may move about quite a bit. To most of us the prospect of a year or more away from home is not one we would chose, particularly for those of us who are married or have been away for long periods lately. There are however consolations while serving in the Far East, we get a very considerable increase in pay.

'When Hostilities Only ratings start to be released, every effort will be made to send them back in time for their release in their proper turn according to their age and service groups. Officers will be appointed to organise vocational training for those men who wish to prepare themselves for shore jobs. Finally I can promise you quite a bit of fun and excitement as well as the hard work. That's about all I can tell you at present.'

The Bishop of Down and Dromore, arrived onboard *Glory* on Sunday 4 March 1945, to consecrate the ship's chapel. The well attended service began at 1000, then, at 1400, the ship was open for a visit by the foremen of Harland & Wolff, accompanied by their wives and girlfriends. Sir Basil and Lady Brooke visited the ship on Monday 19 March, to present a silver cup which would be awarded to winners of inter-departmental competitions. It was Lady Brooke's first visit since the launching.

At last, on Friday 23 March, the special sea dutymen closed up, and HMS *Glory* steamed majestically out into Belfast Lough to carry out anchor trials escorted by HMS *Westcott*, a 1918 'W' class destroyer. The ship returned to her berth for the week-end and left early Monday morning for sea trials, during which *Glory* carried out a 'full speed' test in which she covered a distance of 25.54 nautical miles in an hour, and used 1.63 tons of fuel per mile. The dust of Belfast was well and truly being swept away.

The ship's company now began the process of becoming an efficient fighting unit. There was though still time for enjoyment and some overstepped the mark, like Mr Lewis Davies, Warrant Stores Officer RN who was cautioned by the Captain 'on account of his over indulgence in alcoholic liquor and subsequent misbehaviour on the evening of Wednesday 21 March'. The first aircraft deck landings and take-offs were carried out from *Glory* on Friday 30 March, by Barracuda 11s attached to 837 Squadron. They had reformed at Stretton, (HMS *Blackcap*) as a torpedo reconnaissance squadron on 1 August 1944, but it was not until 4 September before it took delivery of its first two aircraft.

HMS *Glory*

I was born on an Irish sea of eggs and porter

I was born in Belfast, in the MacNiece country,

A child of Harland & Wolff in the iron forest,

My childhood a steel cradle slung from a gantry.

I remember the Queen's Road trams swarming with workers,

The lovely northern voices, the faces of the women,

The plane trees by the City Hall: an Alexanderplatz,

And the sailors coming off shore with silk stockings and linen.

I remember the jokes about sabotage and Dublin,

The noisy jungle of cranes and sheerlegs, the clangour,

The draft in February of a thousand matelots from Devonport,

Surveying anxiously my enormous flight deck and hangar.

I remember the long vista of ships under the quiet mountain,

The signals from Belfast Castle, the usual panic and sea-fever

Before I slid superbly out on the green lough

Leaving the tiny cheering figures on the jetty for ever:

Turning my face from home to the Southern Cross,

A map of crackling stars, and the albatross.

CHARLES CAUSLEY

The squadron moved to Fearn (HMS *Owl*) and increased to 18 aircraft on 1 December 1944. Their first deck landings proved very successful and 837 Squadron began its sea time with high hopes. But during the ensuing weeks extremely flukey weather caused innumerable changes to the flying programme, so that aircraft were ranged and struck down in never ending succession. The 'troops', so many of whom were initiates to carrier life were thrust straight away into frenzied activity. Unceasingly the parade of aeroplanes went on. At an indecently early hour aircrew and flight deck personnel were dragged from bunks and hammocks, their bellies rattling with hunger. The first flush of day however seemed to imbue the flight deck handling party with a new found energy which they speedily dissipated by trundling aircraft headlong down the flight deck.

Propellers were shivered, trailing edges smashed, fuselages dented, such mishaps invariably delayed take-off and happily permitted the persecuted aircrews to disappear like furtive Oliver Twists into the bowels of the ship. But as the anxiety of the first days of Deck Landing Trials wore off, the arrival of *Glory*'s pilots displayed more and more control. Lieut Ted Kirby, 837 Squadron's very own 'Tom Thumb', quite often could not see either 'batsman' or flight deck! This was mainly when parachute packer Sam Sagar had forgotten to inject a judicious quantity of CO_2 into his dinghy, which misfortune produced some very erratic approaches! Then the brightness was dimmed and sadness came to *Glory*. On a navigation exercise from the ship Sub-Lieut Snape RN, Sub-Lieut Pickles RN, and Leading Airman Ryan, did not return.

For three days the weary aircrew of 837 Squadron carried out a continuous search from Ireland to remote and forbidding Hebrides, but to no avail, and *Glory* paused to mourn a gallant crew. Tragedy struck again the following day Wednesday 18 April 1945, when one of the flight deck team, Able Seaman Emerson, was fatally injured when he ran into Lieut McGuire's propeller; Able Seaman Mason was also badly injured in the same accident. The flying programme continued unabated, everyone now only too aware that life on a carrier could be extremely dangerous as well as exciting. With the first part of her work-up finished, HMS *Glory* arrived in the Clyde and on Friday 11 May, secured to her berth in the King George V dock in Glasgow.

2

Japanese Surrender & Repatriation of Allied Prisoners
(June 1945 – December 1945)

Some more aircraft for 837 Squadron were hoisted aboard, followed by the Corsair aircraft of 1831 Squadron. Reformed at Brunswick on 1 November 1944 as a single seater fighter Squadron, 1831 received 18 Corsair 1vs, and sailed to the United Kingdom in HMS *Pursuer* in February 1945. With the arrival of these aircraft, HMS *Glory* was fully equipped with her full operational complement. It drizzled relentlessly on Monday 14 May 1945, as *Glory*, unassumingly slipped her moorings and with little ado glided down the Clyde and passed the boom at 1750 hrs. Straining at the leash outside were the escort destroyers *Icarus*, *Hotspur* and *Escapade*. With a preliminary 'whoop' on their sirens they set the pace and passed the now familiar islands of Great and Little Cumbrae.

Continuing down the Firth of Clyde, the pale evening sun suddenly broke through eerily to pierce the mists of Arran. Pensive looking matelots wistfully gazed shorewards as they took their evening stroll on the flight deck. High above a dozen Barracuda aircraft of 815 Squadron from Machrihanish, were peeling off to carry out a dummy strike on the ship. But 'dummy' was rather a misnomer as the affair became rapidly realistic as they presented their compliments with odious missiles hurled at the ship, resulting in damage to two of *Glory*'s Barracuda aircraft. On 18 May, the ship's company surfaced to consider a sparkling blue sea! 'Land Ho'. To starboard lay darkest Africa, and ahead the Rock of Gibraltar. The ship anchored briefly in the bay, but at noon steamed naked and unescorted along the north African coast.

Exotic names which shone like jewels from the story books: Tangier, Oran, Algiers, Tunis and Bizerta, stimulated the imagination of the ship's company as *Glory* made her way across the Mediterranean Sea, and time burbled pleasantly by. The small volcanic island of Pantelleria, midway between west Sicily and

Tunisia, was passed. On 21 May, after sighting HMS *Duke of York*, *Glory* came to anchor in Marsaloxx, Malta, and met her sister ships *Vengeance* and *Venerable*. The next day the ship steamed further eastwards, *Glory* now in company with the two carriers and the destroyer *Tuscan*. Exactly ten days after leaving the Clyde, the famous city of Alexandria was reached; HMS *Colossus* was in harbour so the 11th Aircraft Carrier Squadron was complete.

Before entering harbour both of *Glory*'s squadrons had flown to Dekheila, the pre-war Alexandria airport. The Admiralty had taken over the airport on 24 May 1939, and as it was the practice for all shore establishments to be given ships' names, Dekheila became HMS *Grebe*. For the next few weeks HMS *Glory* continued to work up using Alexandria as a base. Generally everyone was enjoying the experience of exploring the dens of iniquity in a foreign seaport, some managed to get in a few scrapes whilst ashore, and one rating was brought back by shore patrol on 28 May suffering from a slight knife wound. At sea there was more excitement when a mine was spotted to starboard of *Glory*, and the 'M' class destroyer HMS *Musketeer* was quickly dispatched to sink it, following which flying was resumed, as ever.

Tragedy struck the ship again on Friday 1 June 1945 when a Corsair of 1831 Squadron crashed on landing at 1512, killing Able Seaman Thompson who had been working on the flight deck. His funeral took place the following day, the first burial at sea from *Glory*. All too soon the work up in Alexandria was completed. *Glory* slid along the Suez Canal, the sun shone relentlessly onto the flight deck and like a slow train journey she moved towards the Bitter Lakes, passing on the way an Italian battleship off Fayid. Thursday 5 July was a sweltering tropical day with the ship steaming between Mecca and Port Sudan. Sweat oozed from every pore whether the game was shove half-penny or deck hockey. The iced-water fountains were exhausted time and time again and salt water showers were rigged to try and alleviate the heat. Conditions on the mess-decks were almost unbearable.

On Saturday *Glory* turned into the Gulf of Aden, and the wind freshened until there was a good thirty knots over the deck. The dreadful heat began to moderate and the 'troops' started to quit their breezy beds among the aircraft and returned to their mess-

9

decks. By midnight HMS *Glory* began ever so slowly to roll as she steamed on an easterly course. Thursday 12 July, and the ship reached Ceylon. The coast was low-lying and the hills were hidden by mist, and as *Glory* steamed slowly into calmer waters the air became more humid. The squadrons were flown ashore to Katukurunda, a former RAF base but transferred to the Fleet Air Arm on 15 October 1942, and commissioned as HMS *Ukussa*. The ship meanwhile steamed into Trincomalee, one of the world's finest natural harbours.

The 'comfort' stop in Ceylon was soon over and as *Glory* left the lovely island, 837 Squadron received the news from their Commanding Officer that the squadron was to leave *Glory* and be replaced by 1790 Squadron, who were equipped with Firefly nightfighters. A sea of unusually depressed faces watched the coast of Ceylon disappear as the ship headed for Australia. Happily though, the decision was reversed but not before the 'buzzes' went round the ship that 837 Squadron was changing over to Helldivers, transferring to an American carrier, converting to night fighters, and so on ad infinitum. *Glory* crossed the equator on Saturday 4 August, and everyone was disappointed not to see the reputed 'red bars'.

The time honoured ceremony took place several days later. King Neptune and Queen Amphitrite held their Court in the after lift well, the Clerk of the Rolls immaculately sinister in a snow-white wig, red robe, black patch and evil moustache. Six policemen rushed around blowing their whistles furiously and collecting unwilling victims. The apothecaries and moustached top hatted barbers completed the entourage. One by one the landlubbers and sprogs were shaved by the giant razors and initiated to the mysteries of the deep, making the fearsome looking bears howl continuously for blood. A counter-attack developed by means of well directed hoses from the flight deck, causing confusion among the Royal entourage. Then began a round of dragging people from cabins and messes followed by the routing of the Royal Court of King Neptune.

Accompanied by the destroyer HMS *Wizard*, *Glory* arrived in Fremantle on Friday 10 August, and with their customary alacrity the off duty watch leaped ashore only to be met by mysterious rumours! People were saying that the Japs were suing for peace! But after a voyage across the ocean where the meat and veg had been in short supply and the ship's diet had seemed to consist of corned beef and herrings, *Glory*'s boys were on a quest for steak, egg and chips. Australia was inviting and they were not disappointed, a British pound note bought a schooner of beer and the change included an Aussie pound note! But the rumours continued to circulate around their unbelieving ears. 'It's gen', said someone, 'I've read it in the paper.' But the cynics held out, rapidly downed another Swan lager and moaned something about not recovering from VE Day yet!

At 11 pm the storm broke, the streets of Perth filled with shouting people, rattles churned, paper caps to the fore, the scene was reminiscent of pictures of Piccadilly on Armistice Day 1918. Many of the ship's company were whisked into the homes of the hospitable Western Australians anxious to share the good news, not to mention the beer! It was a confusing night, but there was no confusion about the naval patrols combing the watering holes and other sailors' haunts to order the ship's company back to the ship. On Saturday morning, *Glory* and *Wizard* were back at sea, and on this of all days an exercise was carried between the two ships, with *Wizard* showing very early signs of a hangover. But somehow both ships' companies survived the day.

Wednesday 15 August 1945, it was VJ Day! At 0900 the midnight chimes of Big Ben sounded over the ship's radio equipment, the war quite unbelievably was over! The immediate concern of most of the ship's company was now what lines they could shoot to their children when asked the inevitable 'please daddy what did you do in the war?' The 'mainbrace' was spliced, and in the wardroom a number of notable de-baggings took place, as well as everyone writing on everyone else's collar. On the mess-decks a more sober atmosphere prevailed as Commander Hicks broadcasted his message to the ship's company: 'When this fine ship was built in Belfast all the beautiful brasswork was covered with nasty grey paintwork. This may have to be accepted when there's a war on but that's not how we do it now we are at peace. Working parties will start early tomorrow on scraping, polishing, and burnishing. Be sure to do it properly for if you don't I shall be pointing out the bits you have forgotten.'

The following day flying started very early. Both watches of the flight deck handling party, both watches of the hangar handling party, 837 Squadron, 1831 Squadron, along with Air Maintenance division were called at 0520. At 0550 they fell in on the flight deck, prepared for flying, off covers and securing gear. At 0630, 15 Barracuda aircraft were catapulted off, followed by the 25 Corsairs. After breakfast the aircraft handling party returned all securing gear and chocks to the stowage areas, and then rigged the Ensign and Jack masts before securing the flight deck for entering harbour.

Sydney Heads were in the offing, all *Glory*'s aircraft had flown ashore to the Royal Naval Air Station at Scofields, just north of Sydney. The sun sparkled on the sea and HMS *Wizard* followed *Glory* into harbour. Everyone seemed eager to see some bridge or other! *Glory* passed between two sharp square cliffs, escorted by schools of friendly dolphins who glided smoothly in and out of the water. The hands fell in for entering harbour, including the ship's cat, and the Royal Marines band inevitably began to play. The ship was enveloped in smoke from a tug, and as *Glory* emerged from the haze the famous bridge appeared, but the peculiar thing was, there was no one to welcome them.

They suddenly twigged! It was VJ Day plus one! A ferry mournfully tooted its whistle in passing and a few passengers draped over the rail flapped fish-like hands and grinned pallidly. A sign urged everyone to buy victory bonds, and a flight of RAAF Mosquito aircraft above beat up and down. HMS *Glory* finally berthed at No 3 Wharf, Wooloomooloo, opposite the battleship HMS *Anson*. The Royal Marines band played 'Glory, Glory Hallelujah'. Although the celebrations were well under way there was still plenty left for *Glory*'s liberty men when they streamed ashore that day in Sydney. They had steamed thousands of miles to do battle and the only fighting left was over women and beer! Everyone enjoyed the couple of weeks in Sydney; the dangers of war had passed, but the ship was once again a hot bed of buzzes. Would they stay down-under as part of the British Pacific Fleet? Or would they be sent home quickly?

The rumours were quickly dispelled when *Glory* received orders to proceed for Jacquinot Bay, New Britain, and there to take the surrender of all the Japanese Forces occupying New Guinea, New Ireland, Bougainville and adjacent islands. *Glory* left Sydney on

Saturday 1 September, and the two squadrons were embarked as it was uncertain whether the Japanese would oppose the move. The voyage from Sydney lay through a part of the Pacific essentially uncharted since the days of Captain Cook, and 837 Squadron undertook a novel operation by flying ahead of the ship and spotting several reefs which did not appear on the 1945 Admiralty charts. At 0750 on 6 September the ship was stopped because of excessive vibration, hands were piped to action stations but after a short time were stood down again when it was considered that the vibration was caused through an earthquake and not Japanese action.

Shortly after, *Glory* with the Australian sloops *Hart* and *Amethyst* in company came to anchor off Rabaul. The *Hart* was wallowing idly some six hundred yards to starboard when a motor cutter rounded her stern buffeting its way towards *Glory* in a choppy sea, her cargo of dark uniforms emphasized by the glistening white of the boat's crew. Everyone wondered if the 'snotty' in charge wished to contrive some private vengeance, as each wave flung the cutter into the succeeding trough so that spray enveloped the Japanese delegates. Up the starboard after gangway, their sallow faces half resolute, half apprehensive; uncertain, they shuffled about the deck until a stern Officer of the Watch ushered them towards the Quarter Deck.

Their shoddy ill-fitting green uniforms gave them an untidy appearance which was aggravated by their dirty white Peter Pan collars. Some wore tall jack-boots whose immaculate sheen contrasted unhappily with their baggy clothes. Badges of rank were surprisingly inconspicuous and no decorations were worn, nevertheless each of the party bore a veritable armoury consisting of sword, dagger, rifle and revolver. All except the naval representatives favoured the close fitting peaked Japanese field cap pulled low over the brow, these allied with their arsenal of weapons and shabby briefcases lent them a distinctly comic air, and an involuntary thought of Captain Hook's pirate crew leapt to mind but the severe lips belied the benignity of such a concept.

One noted with surprise the immature appearance of the General's weedy ADC. His anxious eyes displayed concern at each of his leader's infrequent changes of expression and to these

13

he responded with motherly solicitude. The Japanese Naval Commander-in-Chief, an aged Vice-Admiral following close on the heels of the General wore a melancholy face which became distraught when he found no Australian representative of Flag rank awaited him. General Sturdee, GOC 1st Australian Army, dismissed the Admiral's protestations and the bizarre party moved on to the quarterdeck there to be disarmed under the sub-machine guns of the grimly immaculate Royal Marines.

On the flight deck speculation intensified, on the bridge Captain Buzzard was still in workaday khaki. He seemed unmoved by the moment and significance of the impending event, and was pre-occupied by navigation and his tiny wheeling fighters, specks against the pale clouds. On deck, work proceeded half-heartedly as figures in gleaming white began to mingle with the maintenance crews, and the white uniforms predominated as the time for the ceremony drew near. Patient sailors conversed in anxious groups or walked gingerly to avoid spoiling their splendour, the fruit of much dhobying in a fiery washroom. Suddenly, the bugle! Everyone rigid while it sounded, then, raced to fall in by divisions. All pretence of work vanished. The random thoughts were interrupted by the Commander's order, 'shun'!

At attention they could feel the steel of the flight deck hot through their soles, the momentary sunshine set ablaze the white uniforms, and the little procession started slowly towards the table. Now they saw them properly for the first time; in front the Japanese General marched slowly, easily, with not unnoticeable dignity. He was a squat figure with features as inscrutable as butter. Behind him, his aides trailed in an untidy double file with yellow faces fixed and expressionless; only the aged Admiral showed any emotion, and seemed to have difficulty in controlling his sorrow. At the table General Imamura saluted, the gesture was acknowledged, from its scabbard he drew his long curved Samurai sword, with both hands he offered it to the Australian GOC. The latter made no move to take it; again, the sword was offered, then a motion towards the table. Hesitantly, the Jap lay it down.

The long rigmarole proceeded with the reading of the terms of surrender, the reply thereto, the instrument of surrender, first in English then in Japanese. They again became conscious of the hot

14

flight deck, the incessant humming of the cameras and the sweat under their uniforms. Now and again the conqueror of Singapore nodded in agreement with some clause, acquiescent to defeat. Nevertheless, throughout one must be impressed with his dignity, and one sensed rather than saw the considerable emotion which he concealed. Once, his ADC reached up to straighten his cap, one had the sudden horrified mental picture of some junior officer presuming to perform such an office for say, Admiral Fraser!

General Imamura's staff however seemed sullen and almost defiant, and considering it, one was amazed. A brief parley, then the signatures began laboriously. The Japanese painted their names on the three copies of the surrender document, the Admiral was now visibly in tears, but to the ship's company witnessing the ceremony it was all somehow impersonal and unreal. It was impossible to associate the simple ceremony with the savagery and violence of jungle battles, and the horrible atrocities. In this act there appeared to be no finality; some emanation from the clouding skies seemed to hover over the assembly making it a fantasy and inconclusive. Then it was over; aides gathered up the documents, the brushes, the pens, and the small Japanese delegation marched slowly back down the flight deck. The lift descended, the Australian General and his staff departed, and the ship's company was dismissed.

On Wednesday 12 September 1945 HMS *Glory* returned to Sydney, the squadrons disembarked to Scofields, and the ship was then prepared for the mercy trip to repatriate several thousand prisoners of war and internees from the Far East. Double decker bunks were erected in the hangar, sufficient to accommodate several hundred people. Also involved in the same duty were the aircraft carriers HMS *Implacable* and HMS *Formidable*. The first women to serve in a British warship, five nursing sisters and sixteen VADs, lined up on the flight deck of *Implacable* when she left Sydney for Japan on Thursday 13th September. HMS *Glory* left with her complement of nurses on Wednesday 28 September, and headed for the Philippine Islands, to pick up the first batch of prisoners.

On the way the ship called in at Manus, a small island in the Bismark Archipelago, then on to Leyte, arriving at Manila, Luzon,

15

on Monday 8 October. The 1,350 British and Canadian ex-prisoners of war were quickly embarked, including 140 hospital cases, 25 of whom were stretcher bound. The ship left the following day and headed for Pearl Harbour. Among the Canadian service men was Sergeant J. Firmin Legros, who had been imprisoned at the Kawasaki Internment Camp on Kyushu Island, Japan. Sergeant Legros said, 'It must have been one of the better camps.' They had been working in coal mines, the food was bad, but towards the finish they got a fairly humane camp commander. Others had not been so fortunate and had suffered appalling deprivations.

For one soldier, HMS *Glory* was British soil; Sapper William Owens had been a prisoner in a Japanese camp and had contracted advanced tuberculosis as a result of his treatment. He was carried aboard at Manila, 'I'm all right now boys', he told his pals, 'I'm back on British soil again and nothing can happen to me now.' But ship's doctors knew he was dying. The best they could hope was that he would last until he got home. Twice Bill Owens got out of bed and walked a few faltering steps. 'See, I'm getting better', he said. A few days later he died, and at 1700, on Monday 15 October 1945, the ship's engines were stopped and *Glory*'s ship's company said goodbye to Bill Owens as he was buried in the Pacific Ocean off Eniwetok Island.

Replenishment of water and fuel was carried out when the ship reached Pearl Harbour on Saturday 20, and she continued her passage on Sunday morning, arriving at Esquimalt, Vancouver Island on Friday 26 October 1945, at 1035. Lower Deck was cleared at 1315 so that a presentation could be made to the ship's company by the prisoners of war, in recognition of the marvellous treatment they had received on their long journey across the Pacific Ocean. Bert Gibbins, from Dennington, near Woodbridge, Suffolk, was one of the soldiers who remember that trip in *Glory* with much affection. He said, 'We went aboard *Glory* and were taken down to one of the mess decks and arrived just in time for the rum ration, and after having no alcohol for nearly four years just a few sips made us feel quite giddy'. Bert and the rest of the British prisoners went ashore that afternoon.

After a short stop in Canada they continued their journey to

New York, and then boarded the liner Queen Mary for the last stage of their voyage home. Leaving the Lions Gate Bridge astern in the Burrard Inlet, HMS *Glory* approached Vancouver the following day. It was high tide when the ship berthed alongside the Canadian Pacific Railway Pier A-2. The carrier towered above the pier, and a heavy gangway being lowered aboard broke away, crashed to the wharf, bounced – narrowly missing four dockyard mateys – and fell into the water between ship and dock. The men leaped away and no one was hurt, but the mishap delayed disembarkation for about twenty minutes whilst the gangway was fished out of the water.

Two thousand people watched as *Glory* docked, a quiet welcome compared with that given to HMS *Implacable* a fortnight earlier when she had been mobbed and stripped of nearly everything movable by souvenir hunters. One of *Glory*'s officers when told about the *Implacable* incident by a Canadian, just smiled and said, 'We're very glad to be here and will welcome Vancouver people aboard.' A steady stream of people came to the shore from up town Vancouver to see the ship lying in her grey livery under brilliant sunshine, a blue sky, and fast sailing white clouds. When the Mayor of Vancouver, Mr Cornett, made his official visit to *Glory* he had with him his sergeant-at-arms, Mr Alex McKay, who had served aboard the battleship *Glory* during the first world war. On Tuesday 30 October 1945, HMS *Glory* was 'open' to the people of Vancouver.

But the visit was to be orderly – or else. Vancouver City Police, Naval Shore Patrols, and members of *Glory*'s ship's company made elaborate preparations to back up the 'or else'. Plans included one well policed route to the ship, no young boys allowed onboard unless accompanied by an adult, and visitors would be met at the foot of the gangway and escorted through the ship by *Glory* personnel. Sentries were posted all over the ship as a precaution against looting. The first queues began to form at 1130, though the ship was not open until 1330! Despite continual rain, more than 5,000 Vancouver citizens came aboard during the afternoon, and the only souvenir hunters were a few glamour girls in slacks, some of whom managed to get away with some sailors for souvenirs but only as a temporary arrangement.

17

The following day many members of *Glory*'s ship's company were taken on a sight seeing tour of the Fraser Valley. The arrangements were well organised by the secretary of the British Columbia Automobile Association, Mr Frank Bird, and the route for the all-day outing was to Chilliwack, returning via Mission. After a memorable visit the *Glory* steamed away from Vancouver on Monday 5 November, her course set this time for Hong Kong where she picked up a number of displaced persons and took them to Manila, where she arrived on Saturday 24 November. Having landed the displaced persons, *Glory* then embarked a contingent of Dutch Army ex-POWs, who thought they were destined for demobilisation but were instead re-kitted onboard and then taken to Balikpapan, Borneo.

They were then put ashore to fight 'insurgents'. The Dutch soldiers must have been bitterly disappointed at being put back on active service after having received such savage treatment from the Japanese. The Officer-in-Charge wrote a letter to Captain Buzzard:

Dear Captain,

This letter is to express our gratitude for the hospitality all of our Dutch officers, NCO's, and soldiers, have received on board of your ship. We have had many hardships to endure, but on our way home from Japan, via Okinawa, to Manila, and from there to Balikpapan, we have received many kind hospitalities from several people, but the best remembrance will be that of our stay on *Glory* which ship will carry on her name honourably. We have only one complaint to make and that is the following; that our travelling onboard of your ship has only lasted for two and a half days and not longer. Thanks to the kind reception here, everyone of us will remember the crew of the *Glory* till his last day. We hope sincerely that the *Glory* will have a glorious future and you sir and all the other men of the *Glory* will serve happily on her.

Sincerely yours

H.A.W. Scheuer
Captain Artillery
RNEI Army

On Wednesday 28 November 1945, the Dutch soldiers were disembarked at Balikpapan, Borneo, and *Glory* sailed the following day to Tarakan, further down the coast of Borneo, where 1,329 Australian Army and Airforce personnel were embarked and ferried home to Australia, where they would enjoy their first leave for two years. After a few hours stop in Manus, HMS *Glory* arrived back in Sydney Harbour on Wednesday 12 December 1945. Two stowaways were discovered after the ship had secured to the jetty at Wooloomooloo, it was said that they would be returned to their units as they had insufficient points to be de-mobbed!

The Australian troops brought back by *Glory* subscribed £247 in appreciation for what the ship's company had done for them during the voyage home. The ship also brought back two 16ft skiffs and one 12ft skiff made out of Borneo cedar by the Australians on Tarakan. The ship's 'newspaper' was published daily on the voyage, and at the 'dog races' held aboard the totalisator handled £347. Among those returning in *Glory* were Sergeant D.S. Hann, Flying Officer R.D. Taylor, and Flight Sergeant 'Bushy' Fox, all from Victoria. They had taken part in a trek of 362 miles through jungles unknown to white men to reach a Kittyhawk fighter plan which had crashed on patrol in Borneo. There was concern in Sydney at this time about beer supplies, but the problem was overcome in time for the festivities. It is not proposed to list all the activities which the ship's company engaged in during Christmas as enjoyment needs no diarist.

3

Australia, New Zealand & Japan
(January 1946 – July 1946)

The Australian people really put themselves out and their kindness was overwhelming. Within hours of arriving in the country most of the ship's company were being looked after by families, or going on expeditions arranged by various private organisations. The beer was creamy and strong and for a ridiculously low sum you could have a meal of sumptuousness which Britain had not seen for years. It was a truly magnificent Christmas and New Year, but all too soon it was time to blow the dust off the flight deck. On Tuesday 15 January 1946, HMS *Glory* left Sydney and headed for Jervis Bay, where the squadrons were embarked without mishap. The Barracuda aircraft of 837 Squadron had been replaced by Fairey Firefly aircraft, whilst 1831 Squadron retained its Corsairs.

The following Monday, *Glory* escorted by the destroyer *Armada*, departed for Melbourne, meeting the fleet carriers *Indefatigable*, *Implacable*, and the destroyer *Tuscan*, off Point Perpendicular. In the afternoon a rehearsal was held for the 72 aircraft flypast, eight of 837 Squadron Fireflies, and eight of 1831 Squadrons Corsairs were planned to take part. The three aircraft carriers turned into wind at 1330, with HMS *Glory* ahead, and as if an invisible hand held three pieces of string with the aircraft tied on at regular intervals, a 'rabbit out of the hat trick' was performed and the aeroplanes were pulled ever so slowly into the air. The aircraft soon formed up and Lieut-Cmdr 'Buster' Hallet circled, egging up each airgroup on the way. After the rehearsal a couple of runs over the carrier force were followed by some good deck landings.

The following day it was pleasantly cold as the ship turned into the Bass Straits, and 'stand easy' was much appreciated. Weather was too bad to fly off near Melbourne, so the Air Group Commander decided on Wilsons Promontory as the place for the aircraft departures and thereby gave the pilots a quiet 120 miles to settle down. *Glory*'s take off intervals were quite good, 10 second

20

average, and the form up better than the day before. The leaders set course at a thousand feet, not so far above some of the Victoria Ranges, and after much chopping and changing of throttle settings the aircraft reached Melbourne to deliver three rip-roaring passes over the city. One significant occurrence, some types even stopped playing cricket to watch! During the course of *Glory*'s Corsairs landing on after the flypast over Melbourne, one of the Firefly aircraft made a pass at the deck out of turn and was waved off.

The Firefly turned steeply over the flight deck to port and white smoke began to pour from the engine, it was Sub-Lieut A.D. Sanderson RNVR, with Sub-Lieut F.R. Binks RNVR in the rear seat. The aeroplane lost height towards the destroyer *Tuscan* and finally ditched. Both crew were soon picked up by *Tuscan*, and during tea reports came through that everything seemed to be all right. Later in the evening a signal came from *Tuscan* to say though Sub-Lieut Binks was OK, Sub-Lieut 'Sandy' Sanderson, one of 837 Squadron's most cheerful and carefree pilots, had failed completely to respond to artificial respiration. Thus the price for showing the flag.

The next day, Wednesday 23 January 1946, HMS *Indefatigable*, HMS *Implacable* and HMS *Glory* slid their combined 80,000 tons into Station Pier, Port Melbourne. It was the biggest assemblage of Royal Naval tonnage that Melbourne had seen for years. While the carriers moved into berth they were watched by thousands of people lining the bayside in mild summer sunshine. The line of private motor cars – most of them packed with spectators – stretched southwards out of sight. Many of the onlookers on Station Pier were puzzled by the long lines of sailors standing to attention on the flight decks as the carriers docked. Some thought they were being drilled, others that they were receiving shore pay! Not many were aware that the ships' companies were obeying a tradition almost as old as the Royal Navy itself.

Centuries ago 'manning ships' in this way was a sign of friendliness, and demonstrated that there would be no surprise attack; a custom still ceremoniously observed today. After the three carriers were secured, the Funeral Party were mustered and went ashore to the Victoria National Cemetery, Springvale, where Sub-Lieut 'Sandy' Sanderson RNVR was buried with full military honours.

21

Six pilots from his Squadron were pall bearers, a squad of eight ratings from *Glory* fired three volleys over the grave and the naval bugler sounded the 'last post' and 'reveille'. Vice-Admiral Sir Philip Vian attended the funeral. Meanwhile on Station Pier, personnel from all three carriers were practising for the big march past in Melbourne planned for Friday.

Hundreds of women and children clapped and cheered as the Royal Marines and ratings rehearsed their march. Disappointed that they were refused entry to the Pier, they were at least rewarded by the magnificent marching of more than a 1,000 'handpicked' officers and men from the three carriers. After the march practice was over all the matelots got down to writing home.

There was unanimity on two points – quality and scarcity of Melbourne ale, and the amiability of Melbourne girlhood. Writing home to his sister in Northampton, Stoker Derrick Simpson said, 'It's good here. It's nice to have people take an interest in you. The Aussies, who are very nice, especially the girls, stop you in the street and ask about the old country.' Stoker Allan Parmentier, writing to his mother in London, said, 'I would like to stow an Aussie girl away mother, and bring her back to Blighty. I would like to stow away some of the food too.'

At least 250,000 people assembled in Melbourne's streets on Friday 25 January 1946 to see the parade of more than a thousand men of the Royal Navy and Royal Marines. 'The Royal Navy on such occasions is something of which every citizen can be very proud', wrote one newspaper. 'The sight of the men marching as one man, the white gaiters moving up and down with split second precision, the swinging arms and the flash of steel, it is all very emotion raising to the staidest civilian.' Melbourne, with crowded streets and flags flying, gave a cheering welcome to the officers and men of the three carriers and destroyers.

They were cheered heartily as they passed along Swanston, Bourke, and Elizabeth Streets, and showered with confetti from the windows of many high buildings. The salute was taken by Sir Winston Dugan, Governor, who was acompanied at the saluting base in front of the Town Hall by Vice-Admiral Sir Philip Vian, Flag Officer 1st Aircraft Carrier Squadron. At the finishing point in Alexandra Avenue, came the welcome, 'Order Arms'; at last they could all relax. Almost to a man they went through exactly the same motions – they could have still been carrying out orders: each man dived for a cigarette, lit it, and then swung his rifle arm to and fro, creased up his face in a sigh of relief, and began pummelling the arm that had carried the rifle. Then the word went round that the Salvation Army mobile canteen was dispensing refreshments.

At the double the men raced towards it. Out of glass cups they drank lemonade, and grabbed a piece of cake. In a very short time they were on the march again, back to the ships. Those who had time to talk of the parade through mouthfuls of cake said, 'It was a very good show. The people gave us a great hand'. Men from *Tuscan* and *Armada* said, 'We wouldn't have missed this for anything.' On Saturday and Sunday, fifty-thousand people were estimated to have inspected *Glory*, *Implacable* and *Indefatigable*. In addition, at least eighty-thousand other people contented themselves with an outside view of these newest Royal Navy ships. All day packed queues in some cases 600 to 700 yards long, moved yard by yard towards the ships.

There was again an organised plan which made provision for access at one end of the ship and exit at the other. Up until 6 o'clock on Sunday night they were still coming, and the queues showed no signs of diminishing. It was Melbourne's biggest jam for years. Motor cars, taxis, buses, bicycles, all sorts of horse-drawn contraptions, even prams and scooters, were parked along the foreshore for over a mile. Itinerant photographers by the dozen and soft drink and ice cream salesman by the score, all profited by the carrier squadrons visit. Radio music amplified by *Glory* on one side of the pier and *Indefatigable* on the other, kept the crowds in good humour. An air force LAC held a baby for three hours on his slow but non-stop drift from the back of the queue into *Glory* where he was given a cup of tea and cheers. As a 'thank you' to the people of Melbourne a special 'At Home' was held in *Glory* on Thursday 24 January, hosted by Vice-Admiral Sir Philip Vian, and the Commanding Officers and Officers of His Majesty's Ships *Indefatigable*, *Implacable*, *Glory*, *Armada* and *Tuscan*.

The entrance to the party began through the intricate passageways of *Glory*, leading into the after lift-well where the hosts received their guests, numbering about 400, who afterwards passed into the hangar which was decorated with masses of bunting, electric lights, and groupings of flowers, including gladioli and kniphofia, and miniature trees interspersed with those laden with golden mandarins. From this spacious setting a delightful silhouette was presented against the grey skyline of a Corsair fighter plane.

Specially lighted for the occasion, the Corsair lent a dramatic air to the scene. The ship's mascot, a black cat, moved nonchalantly among the guests and their hosts. An impressive moment occurred at sunset when the after lift silently rose. When it descended again, a mixed ship's band standing at attention came into view, and the 'Retreat' was played followed by the buglers sounding 'Sunset'. The ships' companies that visited Melbourne were unanimous in their praise of the city; their welcome lacked nothing, scores of dances, visits, and parties were all splendidly organised. The people had come from their homes and places of work in their hundreds of thousands to contribute to the welcome, and as the three carriers and their destroyer escorts left Port Melbourne on Thursday 31 January 1946, there were many tearful young ladies and sad young men.

The following week found *Glory* experiencing her unluckiest two days since she left the Clyde. At 1500 on Thursday 7 February, a Corsair aircraft of 1831 Squadron ditched in the sea off Noruya, and despite a continuous and unremitting search the pilot could not be found. A Firefly aircraft took off on one of the searches at 1800 crashed into the 'island', and later that same evening further tragedy befell *Glory* when at 2100, Steward Cross, sitting in a 'Clarkat' (aircraft tractor), was pitched into the forward liftwell and crushed to death by the falling tractor. Next day at 0918, a Firefly failed to pull out of his jink on take off and crashed into the sea. The pilot got free but there was no trace of Leading Airman Berry.

Both the Squadrons were flown off the ship to Williamtown on Thursday 14 February. *Glory* then returned to Sydney and embarked a contingent of New Zealand air and naval personnel for passage to Auckland; one of those returning was Wing Commander J. Fleming OBE, RAF, formerly of Wellington, who had left home to enlist seven years earlier. It was during this passage that some of the ship's company were employed in dumping the 'worn out' and 'obsolete' lease-lend Corsair fighters into the sea off the coast of New South Wales. It was rather ironic that on arrival at Princes Wharf, Auckland on Saturday 2 March, the ship's company found they were required to load 24 Corsair aircraft belonging to No 14 Squadron, Royal New Zealand Air Force, for passage to Kure, Japan, where they were to augment the British Commonwealth Occupation Forces.

In addition, 36 vehicles, 400 Squadron personnel and 260 tons of stores consisting of 6,000 packages were also brought abroad. There was just room left on the flight deck for deck hockey and PT! The New Zealand personnel were distributed to messes around the ship and proved popular companions for *Glory*'s ship's company. The Sergeants were accommodated in Petty Officers' No 10 Mess, and to their surprise and delight found they were victualled in for rum ration. They enjoyed the daily ceremony of using the measure and tumbler to dispense an exact tot and most showed their appreciation by tipping a generous quantity back into the fanny.

Whilst *Glory* was in Auckland, the ship's company welcomed the opportunity to go on the sight-seeing tours organised by the Flying Angel Mission. The tours took the men around the outskirts of the city and covered about 40 miles. A football match was also arranged plus the added bonus of being able to watch the first post-war New Zealand and Australian cricket Test Match. Thousands of Aucklanders took advantage of the opportunity to visit *Glory* at Princes Wharf, including the Lord Mayor. An announcement had been made that the ship would be opened to the public at 1400, and a big crowd had gathered by 1330, and as before a steady stream of people passed along a specified route for the three hours that *Glory* was 'open'. The ship left New Zealand on Friday 8 March, and began the sixteen day voyage to Japan.

The time passed pleasantly enough with numerous games of deck hockey and ukkers being played on the way. The New Zealanders also showed their generosity by providing many cases of tinned fruit, cream, jam, honey, cheese and other 'goodies' which raised the standard of cuisine in some messes for a long time. The *Glory* secured to number four buoy in Kure Harbour, at 1740 on Saturday 23 March. Wing Commander de Willimof, Commanding Officer of 14 Squadron, was a hero to his men. They had seen service in Burma, and according to one report his Corsair had flown so low along the main street in Rangoon that holes had been shot into the wings by rifles aimed downwards from buildings occupied by Japanese forces!

Before disembarking he wrote a letter to Captain Buzzard:

Sir,

I would like to take this final opportunity of expressing to you and the ship's company as a whole the very real appreciation we feel of the courtesies and goodwill extended to us during our voyage from New Zealand to Japan. A contract of this nature between two services can so easily be just another job of work and as such can be treated with complete impartiality. We as a Squadron have felt that your entire ship's company have taken a real interest in the well being of 14 Squadron, and have in so many ways crossed the line of sheer duty to afford us the courtesies, facilities, and amenities designed to make our voyage interesting and our presence not merely accepted but welcomed. I would like to be able to feel that this voyage has played a small part towards fostering goodwill and understanding between the people of the homeland and the colonies. I would like you to know that we New Zealanders appreciate the difficult years which the homeland and the empire has yet to face and that we away down under must play our part with our greatest endeavours. I will close by wishing HMS *Glory* a pleasant return journey and good fortune in whatever future duty she is called upon to serve.

Yours sincerely,

Commanding Officer, 14 Squadron, Royal New Zealand Air Force.

Whilst in Japan, some of the ship's company were able to visit Hiroshima and see for themselves the utter devastation caused by the first atomic bomb that fell on this Japanese city and seaport only eight months previously. George Reid, from Edinburgh, remembers vividly the painted slabs of wood outside plots where houses once stood recording the names of the people who once lived there. In Kure at that time there was little to see or do ashore and so for the entertainment of the occupation forces an 'amenity ship' was moored in Kure Harbour. Patrons could get a drink of beer, go to the pictures, see ENSA perform, and make a record to send home to Mum! It certainly did not have the ambience of the Gut, the Sun Sun Cafe in Kowloon, or The Great World in Singapore.

HMS *Glory* steamed out of Kure at 0530 on Friday 29 March, and returned to Sydney once more and the delights of the Fleet Canteen at Wooloomooloo, swimming at Manly and Bondi Beaches, dances at the British Centre in Hyde Park, speedway at Sydney Sports Ground, and perhaps watching Lindsay Hassett's Australian Services XI playing New South Wales at the SCG. It was during April 1946, that strong rumours began circulating the Sydney waterfront that HMS *Glory* was to be handed over to the Royal Australian Navy as a 'mother' ship for the nucleus of an Australian Fleet Air Arm. At the same time the British Press had announced that *Glory* would be given to the French Navy on loan for five years. But stories that the ship would be transferred to the Royal Australian Navy persisted for many weeks.

Rear-Admiral C.H.L. Woodhouse CB, arrived in Sydney on 30 April 1946 as the newly appointed Rear-Admiral Commanding British Pacific Fleet Aircraft Carriers, and would fly his flag in HMS *Glory*. The ship spent a few weeks in the Cook dry dock for some much needed maintenance, the Squadrons meanwhile were still enjoying the comparative luxury of the Air Station at Williamtown. Newcastle, New South Wales, turned out in force to greet *Glory* and her escorting destroyer HMS *Finisterre*, when they arrived on Saturday 15 June 1946 for a two day visit. Both ships were 'open' to visitors, and several of the ships' officers fell for duties conducting tours, or acting as 'commissioned' lift drivers.

They showed what seemed to be the whole population of Newcastle up one lift across the flight deck and down the other lift. The steady stream of people continued for both days and at one time the queues filled every wharf in the area. Labours ceased at sunset with thousands of disappointed citizens being turned away. HMS *Finisterre* reported the loss of everything that was movable! The gunner counted his guns! Back at sea on Monday, and *Glory* received a signal from the battleship HMS *Duke of York*,

From Admiral Fraser:
I have never been fortunate enough to have your company. I am very mindful however of what you have done, the enemy's surrender, the passengers you have carried, the miles you have steamed. You have thus done great work for the British Pacific Fleet and I am very grateful. Message Ends.

The next day, three Firefly aircraft were launched with pilots in the back seats to bring off the last remaining aircraft left at Williamtown. With their return, the ship's company were once again brought face to face with tragedy when Sub-Lieut Jimmy Lawson spun in on his approach over the starboard side, but despite a long and thorough search of the area by HMS *Finisterre*, no trace of Sub-Lieut Lawson was found. Once more everyone felt the loss of a pilot in peace-time so much more than in war. Hardly had the ship's company had time to accept this loss, when on the following day one of the Flight Deck Division on his first time as a chockman, was blown by a Corsair's slipstream into the propeller of another aircraft. Twice death had cast its shadow over the ship and no one was sorry to finish flying at the end of an unhappy and tragic week and return to Sydney for refuelling.

Next port of call was to be the lovely city of Adelaide, capital of South Australia. Exercises on the way demonstrated the use of aircraft carriers to the Australian Navy Board, and on 21 June extensive rehearsals were carried out for the forthcoming demonstrations which consisted of strikes and interception exercises. Early in the morning of Monday 24 June, off Flinders Point, *Glory* received onboard a party of VIPs including, among others, the Minister for Navy and Minister for Trade. *Glory*'s aircraft carried out a live rocket projectile strike in the morning and Lieut Smith provided excitement with incident number one.

His landing was good enough, but his hook after catching a wire snapped! He lost his wheels in the first barrier and came to rest in barrier two, Lieut Smith emerged unhurt followed, if not led, by his observer Sub-Lieut D.S. Williams. Incident number two occurred when returning from a demonstration interception, Sub-Lieut Bennett hit the round-down and finished up facing the crane, quite spectacular, but both occupants lived to tell the tale; so did the cine-photographer who stayed put for a complete record of the prang from start to finish even though the aircraft ended up four feet from his camera! That evening the Australian guests departed, some keen to have a carrier for the Australian Navy and others not so sure. Early on Wednesday morning, the 16th Carrier Air Group flew past over Adelaide in an 'A' formation.

After several runs one of the aircraft was detached to fly low over

Adelaide racecourse to drop press photographs of the exercises which had taken place two days earlier. The photographs greeted *Glory* from the front pages of newspapers as she steamed into Adelaide outer harbour on Thursday 27 June 1946. There was a reception for the ship's officers given by the South Australian Government which started off a round of very successful social activities for the ship's company as a whole. The ship's rugby team had the distinction of beating a South Australian team by sixteen points to nine. On Saturday and Sunday, despite wet weather, over eighty thousand people visited the *Glory*. Wherever *Glory* went in Australia the welcome had been truly magnificent.

4

Ceylon, Hong Kong, Singapore & Bombay (August 1946 – June 1947)

But sadly, *Glory* had other duties to carry out, and on Friday 5 July 1946, she left Australia and made for Ceylon. By some queer trick of fate it was decided not to carry observers in the Firefly aircraft for deck landing practice, when flying commenced on the morning of 12 July. The sea was calm and the weather tropically perfect. Lieut Squires made an approach with no wheels or hook down and his radio seemed duff; a second approach and he had one wheel and one flap down. He climbed and tried to force his landing gear and hook down but with no success, eventually it was decided he should bale out. The Flight Deck Party quickly painted the message on the deck, 'Bale Out'. Some said it should have been 'Bail', but no matter. After several runs he seemed to get the idea and climbed away from the ship all eyes focused on the disappearing aeroplane.

A terrific cheer told those who could not see that a yellow parachute was descending. Already HMS *Finisterre* was making her way fast to the probable spot where he would drop. He was next seen waving as his dinghy took him down the port side and soon he was aboard HMS *Finisterre*. By signal it was learned that his leg was badly smashed but all on *Glory* were thankful.

The period spent in Ceylon could be regarded as one devoted to much change: *Glory* said a fond farewell to 1831 Squadron, they discarded most of their Corsair aircraft and transferred to HMS *Vengeance* for passage home, and disbanded after arrival on 13 August 1946.

Their place on *Glory* was taken by 806 Squadron, who were equipped with Seafire XVs, and remained ashore but carried out deck landing practice when the ship went to sea each week. There were large numbers on *Glory* whose turn for demobilisation had come, and they took passage home in either *Vengeance*, *Indomitable* or *Tyne*. Captain Sir Anthony Wass Buzzard RN had been relieved by Captain W.D. Couchman DSO, OBE, RN, and the new Flag Officer Aircraft Carriers was Rear-Admiral A.R. Bridges, who had taken over from his distinguished predecessor Rear-Admiral C.H.L. Woodhouse. HMS *Glory* was now under completely new management. During the last year the working up of the ship's squadrons had never really got into its stride due to the constant changing of Royal Naval Volunteer Reserve pilots.

From August 1946 onwards the aircrew were all Royal Navy or Royal Navy (A), and it was possible to settle down to progressive training. After embarking the Firefly aircraft of 837 Squadron, and the new 806 Squadron with its brand new Seafires, HMS *Glory* left Trincomalee on Saturday 21 September 1946 and made for Hong Kong. Nobody was sorry to be steaming north. For one thing it was going to be much cooler, and for another, most of the ship's company had exhausted the possibilities of Ceylon as a place of amusement and had become bored with its exuberant lushness and uncomfortable stickiness. A good deal of real naval aviation was done on the way up culminating with an exercise with HMS *Belfast*, the flagship. The ship arrived in Hong Kong on Tuesday 1 October 1946.

The ship's company found Hong Kong a very congenial place after Trincomalee. The Colony seemed to have recovered from Japanese occupation remarkably quickly and the shops were full of exciting things to buy. Socially everyone did very well although nice girls were fairly scarce, but those who were available were very conscious of their worth. There was also one of the finest canteens ever seen by matelots, and there was no shortage of beer and ample food. HMS *Glory* went to sea once a week to give the squadrons deck landing practice and although the weather was often rainy, the temperature remained pleasantly cool and everyone benefited by the change. As with most big ships, when in harbour HMS *Glory* had her round of ceremonial duties to perform and on Friday 25 October the 'Beating Retreat' ceremony was performed.

This is always impressive, but it was particularly so when carried out by the massed bands of the Royal Marines on the flight deck as the carrier swung slowly in the harbour. Despite the threatening weather, a large crowd including His Excellency the Governor of Hong Kong, high ranking officers of all three services and many

31

ladies gathered to witness the display. From the smart appearance of the ship it was obvious that much preparatory work had been done. It was even rumoured that some of the officers had helped with the painting! Just before 1830, the rising of the forward lift which seemingly heralded the arrival of the band, only revealed one small cat, the ship's mascot had arrived!

The ceremony could now begin. And begin it did a few moments later. Simultaneously the two lifts rose, the after one carrying the ceremonial guard, the forward one the bands of HM ships *Belfast*, *Bermuda* and *Glory*, playing as it rose Alford's 'By Land and Sea'. With that precision of which only the Royal Marines are capable, the band marched and countermarched along the crowd-lined flight deck. Led by Drum Major Skippings they continued with 'Quarter Deck', played in both slow and quick time. Then came 'Toc H', played by the drummers of 42 Royal Marines Commando, distinguished throughout by their green berets, to precede the main item – the 'Beating Retreat' itself. So accurate had been the timing of Captain Gosling, Royal Marines, that the bands, conducted by Fleet Bandmaster J.H. Gale, went without pause or delay into the 'Sunset' call as the White Ensign was slowly lowered.

Harmonious and touching was the scene with everyone standing in silence, the dusk quickly closing in, and the band playing the soft melody. 'Reveille' was followed by the hymn 'The day Thou gavest, Lord, is ended', and the National Anthem. The ceremony ended to the strains of that most famous of regimental marches, 'A Life on the Ocean Wave'. A further experience was in store for visitors when the 'island' of *Glory*, and a Firefly aircraft at either end of the flight deck became floodlit. A previous HMS *Glory* had carried the Worcester Regiment as her Marines, (they fought with the ship at Belle Isle) and consequently the ship had the privilege and honour of using its Regimental March, which was also played that evening.

On Monday 4 November *Glory* left for Singapore after embarking the two squadrons, and routine flying exercises were carried out on the voyage. Early in the morning of 18 November the flight deck presented an overcrowded picture, four remaining Corsairs were up forward, then the Walrus, then seven Seafires, with behind them a solid phalanx of ten Firefly aircraft. After briefing, a swarm of pilots manned their aircraft to fly over Singapore only to be overtaken by an impenetrable rainstorn, so the take-off was delayed until this passed. At 1000 the Corsairs were launched and the others followed without incident. An aerial tour of Singapore Island followed, then a 'shoot up' of the town. All the aircraft landed safely at Sembawang, a small green airfield.

Climatically, Trincomalee was less exhausting than Singapore, but living on *Glory* in the tropics was unbelievably uncomfortable for the ship's company. The squadrons were fortunate that at Sembawang everything was done to make them comfortable. The officers were accommodated in a wooden annex to the mess which was a colonial style building reminiscent of the works of Somerset Maugham. The mess itself was a beautiful building of high colonnades like a Greek temple. There were also a swimming pool and tennis courts. To complete the amenities there were small Chinese civilians with gold teeth and frizzy hair who did the dhobying. The ship returned to Hong Kong in December, and remained in the Colony for Christmas, when a large number of the ship's company experienced their first shipboard festivities.

The first few days of flying in 1947 terminated in tragedy. Lieut Dan MacKinnon, a well known pilot and CO of the Corsair Flight, had engine trouble and whilst circling the ship for an emergency landing, ditched on the downwind leg. He was thrown through the windscreen and suffered a dangerous fracture of the frontal portion of his skull. For days he was on the danger list and then made a miraculous recovery, but it was thought he would never see again. To one who loved flying as he did, the disappointment he underwent on learning this fate can well be imagined. Lieut MacKinnon was transferred to HMS *Venerable* for passage home and everyone thought he would stay alive, restored almost by a miracle, but when he reached Trincomalee he became very ill indeed, contracted meningitis and died on Tuesday 4 March.

Then if that were not enough, the CO of 837 Squadron, Lieut-Cmdr Hamilton-Bates RN, and the new Senior Observer, Lieut Mayne, were both killed during a cannon attack on a splash target towed by HMS *Jamaica*. The attack was initiated at an angle which some thought was dangerously steep. A Sub-Lieut flying number two to the CO said that directly after pull-out the wing tips folded and the aircraft turned over and dived into the sea. Nothing was visible but the pock mark of foam such as is left by a bomb. This accident which occurred on Friday 7 March was a great loss to those in *Glory*, and to naval aviation, for Geoff Bates was an inspiring leader and man of intelligence and as a member of *Glory*'s rugby team, was well known to ship's company.

Command of 837 Squadron was temporarily taken over by Lieut A.R. Payne RN (until the arrival of Lieut-Cmdr R.H. Hain RN) with Lieut Angus Turnbull RN as Senior Pilot and Lieut Cliff Gould RN the Senior Observer. The next day *Glory*, in company with HMS *Contest* and HMS *Jamaica*, arrived in the great Indian seaport of Bombay, passed to Britain in 1662 as part of Catherine of Braganza's wedding dowry. As well as carrying out exercises with the Royal Indian Navy at sea, many sporting activities were also undertaken, one of which was volleyball; *Glory* played a very strong Indian Navy team, and lost. Back in Trincomalee on 25 March Petty Officer Bertram Hubbard sailed a dinghy over to see his oppo, Chief Cox'n Bob Mason, in HMS *Finisterre*. The destroyer had been *Glory*'s escort for some time but her commission had ended now and it was time for a farewell tot. She left for the United Kingdom next day with everyone's good wishes.

On 23 April, *Glory* experienced another dreadful flying accident when the last Seafire to land on skidded across the flight deck and hit eighteen year old Able Seaman Taylor, who died of his injuries two hours later and was buried at sea. Then on Tuesday 29 April as Lieut-Cmdr Thurston the CO of 860 Squadron came into land the stern of the ship rose sharply, the Seafire lost its undercarriage and skidded across the flight deck before plunging into the sea over the port side. The aircraft disappeared immediately giving Lieut-Cmdr Thurston no chance to escape.

During the second week in May great excitement was generated when the Fleet held a Regatta in Trincomalee Harbour. The main regatta was between the four big ships – *Glasgow*, *Jamaica*, *Theseus* and *Glory* – but the destroyers *Contest* and *Constance* had their chances as well, together with the shore bases *Bambara* and *Highflyer*. All the racing took place in the forenoons and during the

34

dog watches, and each ship ran a tote. It was *Glasgow* who won the Cup and became 'Cock o' the Fleet', *Glory* sadly was last although the ship's stokers did magnificently to win their race and according to reports the ship's company raked in about £700, with three officers reputed to have made a killing of £300 between them!

Glory left Ceylon on Monday 12 May, and after two days of flying steamed for Singapore where, on Saturday, as the ship approached the naval dockyard, an aircraft flew over and dropped a message onto the flight deck which started a big hunt onboard for opium. Over three hundred pounds of the drug was discovered, said to be worth in the region of £30,000. It had been hidden in Officers' cabins, under the organ in the ship's chapel, in coconut shells and boxes of tea. The Master-at-Arms was apparently offered £500 if he did not search the wardroom pantry where the bulk of the drugs was found. Several of the Chinese messmen and stewards were rounded up and taken ashore when the ship secured alongside at noon. *Glory* went into dry dock the following day for another short overhaul.

While work was being carried out the ship's company were accommodated ashore in the Royal Naval Barracks. The squadrons had been disembarked to Sembawang where they carried out exercises with the Royal Air Force at Tengah, and also with the Army, so a great deal of useful flying had been done. Wednesday 2 June was a sad day for 837 Squadron, for they buried young LAM (E) Walmsley, who died in the Military Hospital at Singapore, after a lorry in which he had been going ashore with other liberty men turned over on a sharp bend. They had lost a most popular and hardworking lad. At the sale of his kit his shipmates with great generosity raised over £200, which was sent to his parents but everyone felt that no amount of money would compensate them for the loss of their son. Naturally everyone aboard was now looking forward to the Australian visit.

The ship went to sea on Thursday 19 June to fly on the two squadrons, for the remainder of that day and part of the next carrier drill was carried out as the aircrews got back into practice. On Friday, *Glory* returned to Singapore dockyard and commenced loading stores and the spare unserviceable aircraft which had been brought from Sembawang. Everything was stowed away by noon on Saturday. A final cocktail party was given by the ship's officers on the quarter deck to celebrate leaving Singapore, and the ship's second voyage to Australia began at 0600 Monday 23 June 1947. The 'crossing the line' ceremony took place next day and there were many new initiates.

5

Australia Revisited & Voyage Home
(July 1947 – October 1947)

All onboard thought it would be a leisurely cruise, but the Admiral in HMS *Theseus* had other ideas and had arranged a programme which filled nearly every day. The weather was particularly foul and *Glory* only managed to fly on two days. In spite of the motion of the ship, which really became alarming at times, the landings all held and the ship had not a single casualty either on deck or in the hangar caused by aircraft breaking loose. Early on Thursday 3 July, *Glory* flew off all available aircraft to 'beat up' Adelaide and announce the ship's arrival. There were eleven Firefly and ten Seafire aircraft, and as it became light the group formed up and made for the city where the lights were being extinguished one by one, and the citizens were setting about the wearisome business of getting up and cooking breakfast.

After half an hour of noisy flying to and fro, the inhabitants of Adelaide could no longer be in any doubt of what was going on and where it came from. On returning to *Glory*, Lieut Threlfall RN had hydraulic trouble and after a long spell of pumping managed to get everything down except his tail wheel, but as the deck park was full by then, and the time to make the tide was short, he was sent to Parafield. The welcome was as usual magnificent, with dances, parties, trips, tickets for theatres and cinemas, and free public transport. On Monday, a reception was held in HMS *Glory*. There were some 500 official and private guests. A wooden bridge spanned a pond where the after lift-well usually is, gum branches adorned the bulkheads and the hangar was gay with bunting and flags.

Halfway through the evening the noise of conversation, encouraged and loosened by alcohol, was enough to remind a stranger of an aviary gone crazy, and there were lots of pretty girls. On Wednesday 9 July, the Lieutenant Governor and many distinguished citizens came aboard to see *Glory* at work and to send her on

37

her way. In fine weather *Glory*'s aircraft did a form up and a rocket attack on a splash target, which, considering the hangover most pilots had was fairly accurate and in any case spectacular. Afterwards the aircraft flew over the town, once up and once down, so that all *Glory*'s friends could say goodbye. At 1300 the guests left and the ship set course for Melbourne. A mass flypast had not been arranged, so on nearing Melbourne on Friday, 837 Squadron droned over the city a couple of times whilst 806 Squadron were doing an aerobatic turn for the benefit of the Royal Australian Air Force, after which all the aircraft landed at Point Cook.

The ship's arrival at Station Pier, was very subdued when compared with *Glory*'s previous visit to the city. HMS *Theseus* had arrived an hour earlier and was also at Station Pier, whilst the escorting destroyers *Cockade* and *Contest* secured at No 1 South Wharf. *Theseus* was flying the flag of Rear-Admiral G.E. Creasy, now Flag Officer (Air) Far East. All four ships were open to the people of Melbourne on Sunday, thousands came along and there were big queues all day. Showing at one cinema that week were Katherine Hepburn, Spencer Tracey and Melvyn Douglas, in 'The Sea of Grass'.

On Tuesday, the Carrier Squadron went to sea for some demonstration flying. Included in the programme was a flight by 'Montague' a Tiger Moth built in 1934, and belonging to Captain R.K. Dickson RN of HMS *Theseus*. It was flown on this occasion by Commander Robert Everett, son of Australia's first Navy Board Member. Commander Everett climbed aboard and the flight deck was cleared. 'Montague' bustled forward, appeared to hesitate – and suddenly, after a run of less than twenty feet was passing the island 30ft above the flight deck. Meanwhile aboard *Glory*, another rocket attack on a splash target; the shooting was bad although the uninitiated are always impressed by the moderately close backing up of aircraft in the circuit and the landings. The Royal Australian Air Force went away thoughtfully, saying the whole business appeared a bit dicey to them. How very right they were!

Among those entertained onboard during the nine day visit were some of the 'Miss Australia' entrants, Dawn Latter, Joyce Young, Dawn Ingram, and Phyllis Irvine. A group of country schoolchildren were also welcomed aboard who had left their home town of Benalla at 5 a.m. for the trip to Port Melbourne. During the visit, over a thousand men from the four ships marched through the city to express their thanks for the tremendous time the citizens of Melbourne had shown them. The visit had simply flown by. Finally, after a boozy, romantic and practically sleepless week, *Glory* got ready to sail for exercises with the Royal Australian Navy, and prepare for the visit to Sydney.

For some reason it was decided to do a day's flying on Sunday 20 July, thus completely destroying the week-end run ashore. Making due allowances for the unbridled debauchery, (the official ship's party had been held the night before) the landings weren't too bad, a fact which made the tragedies which followed not so easy to account for. The wind began to drop and *Glory*'s large range was reduced by flying off only the Seafires. Whilst this was going on people on the flight deck were watching the aircraft from *Theseus* forming up, when out of a gaggle in the sky dropped something which spiralled down and disappeared into the water with a fountain of white to mark the spot.

Two of their Firefly aircraft had collided and four aircrew were killed, they were: Lieut-Cmdr (P) Nathaniel Martin Hearle RN, Lieut (P) Raymond Thomas Walker DSC, RN, Lieut (A) Kenneth Alfred Sellars RN, and Chief Petty Officer William Lovatt. One of their Seafires then made an erratic landing at 1445, slewed round narrowly missing the batsman, and then hit and fatally injured Ordinary Seaman Anthony E. Timmons, who was in the flight deck walkway. *Glory*'s turn came later when the Seafires returned. Lieut-Cmdr Waller RN the CO of 806 Squadron made his pass at the deck too fast, bounced over the barriers and crashed into the aircraft parked forward. Air Mechanic Sadler was seriously injured and was transferred to HMS *Contest*, who rushed him to Port Melbourne where an ambulance was waiting to take him to Heidelberg Military Hospital, but he was dead on arrival.

Petty Officer Primrose was also injured in the crash which left the flight deck looking as if a Kamikaze attack had taken place. Three Seafires and a Firefly were complete write-offs, and three other Firefly aircraft were damaged seriously enough to engage the Maintenance Unit for some time. The pilot was unhurt and everyone considered it lucky that many more flight deck personnel had not been killed. It seemed clear that on a cruise like that it was impossible to combine serious flying with the strenuous social activities which are inescapable. The alternative solution would have been to stop the leave of aircrews 24 hours before they got airborne or to abandon flying until the ship had been at sea long enough to get some rest. It was particularly unfortunate that the accidents occurred when the two carriers were full of Australian service and civilian guests, including many press representatives.

The Australians had bought two aircraft carriers and must have considered their purchase with mixed feelings. After taking part in a mock 'battle' with units of the Royal Australian Navy, *Glory* entered Sydney Heads in company with the cruiser *Australia*, the destroyers *Cockade*, *Contest*, *Bataan*, *Arunta*, *Warramunga*, and the frigates *Shoalhaven* and *Murchison*. No aircraft were disembarked when *Glory* arrived in Sydney on Thursday 24 July, but the planes flew over the town and the crews were able to admire the lush garden landscapes, the terraced red roofed houses, and the numerous bays of the harbour area. Everyone who could be spared cleared the lower deck when the ship entered harbour.

For once the hour spent lining the flight deck did not become boring for there were many things to engage their attention, and they could fill their eyes with the glories of the place. As *Glory* and *Theseus* made fast alongside the wharves at Wooloomooloo there were two early eager visitors, Mrs Betty Moore and Mrs Pat King, both of St Kilda, Melbourne, to spend a few days with their husbands, Air Mechanic Tony Moore and Air Mechanic Ernest King. Mrs B. Hubbard of Yowie Bay, Port Hacking, was another early arrival, she was there to see her nephew, Petty Officer Bertram Hubbard from Norfolk. She said this would be her last opportunity to see him before he returned to England. Six members of the ship's company had a special treat in store for them later in the week when they were guests of six nurses from Prince Albert Hospital, who had made a blind date with them a week earlier.

The nurses' invitation was one of many offers of hospitality made to the British Centre in Sydney, by city residents anxious to make the men's stay memorable. Messengers from the Sydney GPO waited for the ships to secure; the messengers, two girls and a boy were welcomed by the men who leaned anxiously over the ship's rail, and carried big bundles of telegrams from Australian friends and relatives, including hundreds of offers of accommodation. Travel was free on city trains, trams, buses and ferries. The ships' companies of *Glory*, *Contest* and *Cockade* each had four days continuous shore leave, whilst those on *Theseus* had only one day because they were needed to paint ship!

The Sydney Herald launched a Fleet Food Fund, which was so successful that each member of the fleet received a food parcel to take home to the United Kingdom. The ship's company held a dance, and the Carrier Squadron held a combined one for all officers on the night before sailing for Brisbane. As usual everyone drank to excess, had no sleep, and went about next day wondering why they persisted in doing such things. But *Glory* at least returned some of the hospitality which had been overwhelming the ship's company during the past few weeks. On Thursday 5 August 1947, *Glory*'s ship's company waved goodbye to all the weeping families flimsily secured to the jetty by long coloured streamers, the ship broke her Sydney bonds and moved slowly into the harbour. On a golden evening *Glory* slipped out of Sydney Heads and steamed for Brisbane.

The following day flying was attempted, but so many appalling things happened in the course of it that the Admiral decided that no more should take place from either *Theseus* or *Glory* until a period of intensive training could be done ashore, which meant *Theseus* must disembark her Air Group. That unlucky ship lost four Fireflies in as many minutes. A pilot hurdled the barrier and plunged into the sea sweeping another aircraft in with it and damaging a third. Then another on landing bounced as high as the bridge, hit it, and cartwheeled into the water. In spite of this ghastly performance all the aircrew were picked up and only one was injured, however one rating working on the flight deck was killed.

Whilst all this was going on Lieut Bowen made an unusually uncontrolled pass at *Glory*'s flight deck, bounced high and wide, and ripped into a barrier: a write off! The hangar was now filled with badly damaged aircraft and 837 Squadron's six remaining serviceable Firefly aircraft were to be handed over to *Theseus* in

Brisbane, leaving *Glory* with little but heaps of tortured metal to resuscitate on the way home. The Squadron wanted to make a gesture and deliver the aircraft to *Theseus* by air, but the Admiral would not allow it to be done. Afraid of losing some more? Or just concerned about the feelings of 812 Squadron? The two carriers arrived in Brisbane on Friday 8 August, all the hands thought it would be a dull place for they had not forgotten their memorable days in Sydney, but many changed their minds after they had encountered the unrivalled kindness of the Queenslanders.

It was unfortunate perhaps that many matelots were absolutely broke after the stupendous stay in Sydney, but dances and trips were arranged, and those who did go ashore were made welcome. There was a march through the streets of Brisbane on Tuesday which assembled at Kemp Place, and then proceeded to City Hall, where the Governor General Mr McKell, took the salute at 1030. Afterwards the men were entertained by Brisbane dignitaries in City Hall. The ship was open to visitors and once more thousands flocked to Bretts Wharf for a last 'open' day. Time marched on and at last Monday 18 August came, and with it *Glory*'s farewell to Australia.

Parting is such sweet sorrow. Alas it seems to figure prominently in the lives of sailors. HMS *Glory* pulled out slowly from Bretts Wharf, flying her long paying off pennant. Outside the harbour the *Theseus*, *Australia*, *Hobart* and four destroyers steamed past and gave three cheers as the ship headed for Singapore. Whilst in Australia, the CO of 837 Squadron, Lieut-Cmdr Hain, a confirmed and greying bachelor, got himself engaged to a nice air hostess of TAA, and Lieut Jarvis, the squadron's favourite 'dypsomaniac' did the same thing with a brown-eyed girl from Brisbane. For the voyage home a comprehensive training programme had been arranged for the officers; at 0630 they did PT on the flight deck, and after breakfast lectures on seamanship and navigation were given, interspersed with arms drill to relieve the monotony.

The intention was that all short service officers should know more about the purely naval side of their trade than they had hitherto learned, in view of the fact that some might well be retained. A similar scheme was projected for the ratings who were staying in the service. It was a good idea, for a long sea passage with no flying was dreadfully boring. The ship arrived in Singapore on Saturday 30 August, and there was an impressive ceremony during the stay when the Regimental Colours of the 1st Queens Own Regiment were embarked for passage to the United Kingdom. Then the long journey home continued through the lush green of the Straits of Malacca.

Onward past the Nicobar Islands, and across the Indian Ocean to Trincomalee. In the Gulf of Aden on Sunday 14 September, with a thirty knot wind and a fairly rough sea to contend with, the Army and Air Force personnel had a job trying to keep their feet during Captain's divisions. Passage through the Suez Canal was a rather tense affair owing to the friction between Arabs and Jews. British ships had been attacked, and during the whole journey along the Canal the ship was closed up at 'action stations', and whilst anchored overnight in the Bitter Lake lights were rigged to deter divers, and armed ship's boats patrolled all night. Finally, the ship passed the monument and statue to Ferdinand de Lesseps, and out into the Mediterranean Sea.

Everyone was able to get ashore in Malta and Gibraltar, to purchase presents for families back home. At last HMS *Glory* arrived back in Plymouth Sound on Monday 6 October 1947, and anchored at 1000. The Army, Air Force, and thirty Royal Navy ratings who had been brought back under escort were put ashore, and after just three hours the ship left for Glasgow, arriving the next day. During the next few days, stores, aircraft, food parcels and aviation spirit were unloaded, after which *Glory* sailed for Devonport Dockyard, arriving on Tuesday 14 October. The ship's first commission had been carried out with distinction. Although no war action had been experienced the expenditure in lives was not light and sixteen members of the ship's company lost their lives. *Glory* remained in the West Country for the next year and a half.

6

Mediterranean Commission
(October 1949 – December 1950)

Whilst in Devonport the *Glory* was used to accommodate the ship's company of Australia's first aircraft carrier, HMAS *Sydney*, which would be ready in November 1948. *Glory* underwent a refit in mid-January until mid-June 1949. Work involved re-building and modifying the bridge and having her mix of single bofors and pom-pom guns replaced by sixteen single 40mm guns. The flight deck arrester gear was also up-dated. HMS *Glory* began her second commission on Monday 24 October 1949, when she left Plymouth to join the Mediterranean Fleet and to embark the 14th Carrier Air Group who at that time were serving in HMS *Ocean*. After a weekend in Portsmouth, the ship proceeded to Glasgow to embark aircraft spares, stores and aviation spirit.

On Wednesday 2 November, *Glory* set off down the Firth of Clyde and headed for Malta. It was an unusually placid voyage across the Bay of Biscay, passing Cape St Vincent where the mighty Spanish fleet had been defeated by the Royal Navy in 1797; onward to Cape Trafalgar, Nelson's greatest triumph; then sailing through the Straits of Gibraltar on a glorious morning with the sun rising behind the 'Rock' and big shadows spreading over the sierras of Spain to port, and the rugged bleak mountains of north Africa to starboard. *Glory* secured to a buoy in Grand Harbour Friday 11 November after the Firefly aircraft of 812 Squadron had flown aboard. There were also four Firefly night fighters known as 'Black Flight' which were fitted with wooden propellers which could shatter into a thousand jagged arrows after a heavy landing.

A ghastly start to the commission occurred on Tuesday afternoon 15 November, when the aircraft piloted by Lieut Bernard Jackson RN crashed into the sea on the starboard side and he was killed. At the end of November, *Glory* went into the floating dry dock in Grand Harbour for cleaning and painting below the water line. When this task was completed the ship secured alongside the wharf at Corradino and a great frenzy of scrubbing, painting and polishing began in preparation for the visit of HRH Princess Elizabeth. Everyone was involved from the most junior rating to the Captain, and a tremendous panic was caused when the side-party's catamaran (essential for painting the black line around the ship at waterlevel) went missing. But life goes on and Thursday 8 December was pay-day, and lower deck was cleared for general payment in the hangar.

Orderly queues of matelots formed up to receive the outrageous sum of about two pounds ten shillings which was placed carefully in the middle of their proffered caps by the 'pay-bob'. By 1830, *Glory* was strangely silent as three-quarters of the ship's company had gone ashore for the evening. A gash-chute sentry sat forlornly on a bollard with a Wills Woodbine clamped in his lips and reading the inevitable Hank Janson story. Many, of course, made for the dubious delights of the famous 'Gut' but the more cultured went to see some of the island's historic and beautiful spots such as the wonderful domed church at Mosta, the old capital of Mdina, or St Paul's catacombs at Rabat. Whatever the choice, Malta was a good run ashore.

From the early hours of Wednesday 14 December 1949, tremendous activity took place in order to put the finishing touches to the *Glory*'s already pristine appearance. The ship was dressed overall with a line of bunting 690ft long from the starboard bow to the starboard after round-down. A big Union flag flew from the jack-stay for'd whilst a huge White Ensign was stretched horizontal by the breeze down aft. Her Royal Highness the Princess Elizabeth, accompanied by Admiral Sir Arthur Power, Commander-in-Chief, was to visit *Glory* as part of the King's birthday celebrations. It was a memorable and unique occasion. The Princess embarked in the Commander-in-Chief's barge just before 1100 at the Torpedo Depot, Msida; Her Royal Highness was escorted by Rear-Admiral H.W.U. McCall, Flag Officer Destroyers, on the route around the berthed destroyers in Marsamxett Harbour.

The first ship to cheer the Princess was the Reserve Fleet's HMS *Lofoten*. Then the Royal Fleet Auxiliaries gave their welcome. When the barge pased near the steps of HMS *Phoenicia*, cheering parties, including Wrens, sent an impressive cheer over the water. Her Royal Highness smilingly waved in reply to the cheering from the ships' companies lining the sides as she passed down the lanes between destroyers.

On the bridge of HMS *Chequers*, the Duke of Edinburgh raised his cap and cheered with the rest of the ship's company as Princess Elizabeth passed. The trim green barge hove to near the submarine depot ship HMS *Forth*, and the National Anthem was heard from the great grey ship.

Concluding the visit to the Fleet units berthed in Marsamxett, Lazzaretto and Pieta, the Commander-in-Chief's barge then sailed round St Elmo and arrived in Grand Harbour where the escort was handed over to two boats from the First Cruiser Squadron. As the barge, its Standard whipping out in the stiff breeze, passed the war tested bastion of HMS *St. Angelo*, the saluting guns aboard *Glory* roared out their Royal Salute. The barge swung alongside the over-hanging bulk of the carrier and the Princess boarded *Glory*, her Personal Standard streaming out from the masthead as she did so. The Royal visitor was escorted to the after lift which then ascended to the flight deck. On the lift with her were Admiral Power, Vice-Admiral the Hon Cyril E. Douglas-Tennant, Flag Officer (Air), Lieut-Gen Browning, Comptroller of the Princess's Household, and the Flag Lieutenants.

Once on the flight deck a Royal Marines band formed of detachments from *Glory* and the cruisers *Liverpool* and *Kenya*, played the opening bars of the National Anthem. The Royal Marines guard presented arms, and the Princess inspected the lined flight deck, manned by representative units of Mediterranean Fleet warships. Over 1,000 officers and ratings were inspected by Her Royal Highness, who stopped for a few words with a number of personnel. At noon the booming of a twenty-one gun salute rolled across the ruffled waters of Grand Harbour. Facing aft in the direction of the cruisers *Euryalus*, *Liverpool* and *Phoebe*, wreathed in smoke from their guns, the Princess stood alone.

The last echoes of the salute died away, and Her Royal Highness left the flight deck and embarked in the barge which had been laying off during the formal visit. The barge, escorted by ships' boats, veered away to a position midway between the cruisers and

Glory. At a given signal a snowy mass of caps was lifted high and three thunderous cheers from the officers and ratings of *Glory*, *Phoebe*, *Liverpool* and *Euryalus* swept round the Grand Harbour. Admiral Sir Arthur Power sent the following signal to the Mediterranean Fleet: I have been commanded by His Majesty the King through Her Royal Highness the Princess Elizabeth to '*Splice The Mainbrace*'. The extra issue of rum is to be made tomorrow, Thursday 15 December 1949. From the shore thousands of people had been able to see the pageantry unfolding in the sunshine.

The following week it was back to sea for more flying, one Sea Fury coming to grief when its undercarriage folded up causing the propeller to bend into the most grotesque shapes. Carol services were held in the ship's chapel, and on the quarterdeck; in the messdecks carols were being sung, mostly out of tune. Christmas Day was a most enjoyable time, there was turkey, pork and Christmas cake. Before dinner everything was laid out on the mess tables, nuts, fruits, sweets and cigarettes. The Admiral came round the ship with his wife, also the Captain and his wife, the Commander, Commander Air, and a score of other officers, which had the effect of making everyone feel like orphans at a benefit treat given by the local gentry. It was nevertheless a very jolly time.

All too quickly the festivities were over, and it was back to sea. The Carrier Air Group was embarked and preparations began for the Mediterranean Fleet's spring cruise, there was an atmosphere of excitement onboard as departure for 'foreign' parts drew closer. So ended 1949.

Tragedy struck on Wednesday 4 January, when a Firefly of 812 Squadron plunged into the sea approximately two hundred yards to starboard of *Glory*. The pilot, Lieut A.W. Turney RN and his observer Lieut T.O. Brigstocke RN were both killed. The accident happened at about 1540, and shortly after the body of Lieut Turney was taken from the water by boat from the Royal Indian Navy destroyer *Ranjit*, and landed at Bighi Naval Hospital later in the afternoon. No trace could be found of Lieut Brigstocke, who came from Walmer, Kent, and was married on 26 November 1949 to Lilian Pace, the elder daughter of Captain and Mrs Pace of Sliema.

The funeral of Lieut Turney took place on Saturday 7 January.

The cortege assembled at Fort San Rocco shortly before 1000, and proceeded to Bighi Capuchin Cemetery, the coffin being borne on a gun carriage. At the cemetery the service was conducted by the Reverend C. Birtle RN, chaplain of HMS *Glory*. Lieut Turney came from Tulse Hill, London.

HMS *Glory* arrived in the lovely Bay of Naples on Thursday 25 January. Visits were made to the ruins of Pompeii laying at the base of the slumbering Vesuvius, and there were boat trips to the Isle of Capri and excursions to Rome. Others preferred to stay in Naples and admire some of its outstanding buildings, the Royal Palace, San Carlo Opera House, or Castel Nuovo. The seedier side of the city was also glimpsed during the stay, grog shops, clubs, brothels, and the appalling poverty. The visit was marred by a dreadful accident on Saturday when a car skidded off the jetty at 0530, and Lieut R. Hill RN was drowned. The car was recovered later in the morning and his body was brought onboard when the colours were half-masted. Lieut Hill was buried in Malta on 2 February 1950. HMS *Glory* made history as the first vessel of her class to enter the port of Tripoli on Friday 3 February. Now the Flagship of Vice-Admiral Sir Cyril E. Douglas-Tennant KCB RN, *Glory* was accompanied by the Battle-class destroyer *Vigo*.

Soon after the arrival the Vice-Admiral went ashore to pay an official call on His Excellency, the Chief Administrator, Mr T.R. Brackley CBE. The call was later returned and Mr Brackley was accorded a seventeen gun salute. A full programme of social activities took place and the Army organised a week-end trip to the magnificent archaeological site at Sabratha. Some of the hands from *Glory* were none too impressed with the catering arrangements which seemed to consist of huge slices of bread and jam washed down with dark brown tea. Whilst on passage from Naples to Tripoli, *Glory*'s aircraft had been taking part with the Army and Royal Air Force in Exercise Bounder 2, and during the exercise *Glory*'s aircraft managed to get the better of the Army.

In one particular phase of the exercise, leaflets were dropped around several suspicious tents. The troops couldn't resist it and turned out to gather them up, during which they were promptly strafed from a low level! The leaflets caused a ripple of amusement as they read: 'All the nice girls love a sailor', and made an appeal for Regular Army recruits! *Glory* returned to Malta on 7 February, and remained there until the second stage of the spring cruise got under way in March, when in company with HMS *Chequers* and HMS *Solebay*, *Glory* arrived in Palmas Bay, Sardinia, on the 2 March. Sardinia, with its wild and rugged mountains was a welcome change from the flat brown landscape of Malta. After leaving Palmas Bay, *Glory* took part in exercises with the carriers HMS *Implacable* and HMS *Vengeance*.

Glory's aircraft set some very impressive deck landing times on Tuesday 7 March:

HMS *Glory*	19 Aircraft in 14¼ minutes
HMS *Vengeance*	14 Aircraft in 18 minutes
HMS *Implacable*	18 Aircraft in 22 minutes

Signal to *Glory* from A/C 3 'That was pretty to watch'.

From a peasant economy in Sardinia to a glimpse of high living and equally high prices, the next call was to Golfe Juan on the French Riviera. Here, many of the ship's company visited the fashionable and popular resort of Cannes, whilst others sampled the delights of the 'Battle of Flowers' in the town of Nice; trips were also arranged to the perfume manufacturers in Grasse, and some ventured down to Monte Carlo.

On Saturday 11 March 1950, *Glory*'s ancient amphibious bi-plane, the Sea Otter, made some courtesy calls along the Riviera, dropping in at Nice, San Remo and Monte Carlo. The next day the ship was 'open' to the sophisticated citizens of the Cote d'Azur, who, judging by the numbers that came aboard, seemed to enjoy the experince. Back to sea on Monday morning and three more days of unremitting flying, until 0930 on Thursday, when *Glory* arrived at the north African seaport of Algiers. The capital of Algeria, the city is situated on the narrow coastal plain between the Atlas Mountains and the Mediterranean; the old town is domina-ted by the Kasbah, the Palace and Prison of the former Turkish rulers, and contrasts with the European style of the French built new town.

Algiers was a fascinating place to visit not least perhaps because it retains vestiges of its piratical past, but it also knew how to cater to the needs of the 1950s matelot! The spring cruise was drawing to a close. After more flying exercises with units of the Home Fleet, HMS *Glory* escorted by the destroyer HMS *Barrosa*, arrived in the rocky fortress of Gibraltar, and secured astern of HMS *Implacable* on the South Mole on Wednesday 22 March. The town was packed to capacity as both fleets arrived to do 'battle', mostly on the sports fields but occasionally in the bars and clubs! The shopkeepers did a roaring trade as jolly jack bought table cloths, cushion covers, fancy dressing gowns and other exotic presents to take home to Mums, wives and girlfriends.

HMS *Glory* arrived back in Malta on Friday 31 March, the 14th Carrier Air Group was disembarked to the Royal Naval Air Station at Hal-Far (HMS *Falcon*), and the ship secured in Grand Harbour. During the next two months the routine was to go to sea for flying in the week, anchoring each evening in Marsaloxx, with the weekends spent back in Grand Harbour. Twenty-one ratings joined the ship on 11 May, having been given passage from Liverpool on the Troopship *Empress of Australia*. They would replace the National Service men who were due for de-mob. On Wednesday 24 May, a party of Boy Scouts and Wrens were taken to sea so that they could witness a day's flying. Disaster struck just before lunch when a Sea Fury piloted by Lieut Mudford RN landed heavily, skidded to port and fell into the sea; tragically killing the pilot.

A mysterious fire occurred on the Flag Deck the following week, and although it was extinguished within ten minutes one rating was transferred to Bighi Hospital suffering from smoke inhalation; fortunately he was released fit next day. Other ships in harbour and preparing for the next cruise included *Pelican*, *Armada*, *Reclaim*, *Dalrymple*, *Vigo*, *Gravelines*, *Phoebe*, *Surprise*, *Ceylon*, *Gambia*, *Chivalrous*, *Eurylaus*, and *Saintes*. The Mediterranean Fleet's summer cruise began in June, and combined visits to eastern Mediterranean countries with naval and air exercises operating with Greek and Turkish naval and army units. Landing parties of Royal Marines and ratings from *Glory* were also planned to take part.

The first port of call was at Corfu, the second largest of the Ionian Islands, and *Glory* came to anchor on Wednesday 14 June

1950. Extensive manoeuvres with the Greek Navy took place over the next few days, but then on Monday 19 June two men were very seriously injured in a mishap on the flight deck. The accident, off Corfu, occurred when a Firefly aircraft bounced on making his deck landing, missed all the arrester wires, cleared both barriers, and collided with other aircraft parked for'd. Petty Officer Thomas Turnbull, of Gillingham, Kent, and Naval Airman William Manley, of Elderslie, Renfrewshire, were so badly injured that *Glory* returned to Malta where the two men could be transferred to the Royal Naval Hospital at Bighi.

Replacement aircraft were obtained from Hal-Far, and the ship resumed the cruise with a visit to the island of Skiathos, in the Aegean Sea. On Saturday 1 July *Glory* arrived in Piraeus, the port of Athens, that renowned city of antiquity. A great many cultural visits were arranged during the week to the many tourist attractions, particularly to the Acropolis, which dominates the city, and the other architectural remains of ancient Greece, such as the Parthenon, the Erechtheum, and the Temple of Athena Nike. For others there were all the attractions and delights of the modern Greek capital, with enough bars, clubs and cabarets to satisfy the largest British naval force to visit the city for some years. During the cruise the pipe, 'Hands to the seaplane stations', was a very familiar one on *Glory*, as the Sea Otter, that relic of naval aviation, was used for collecting mail for the fleet.

The Fleet Regatta was another feature of the cruise. On Saturday 8 July, all the ships came to anchor off Marmarice, Turkey, and competition began in earnest to obtain one of the many cups on offer to the winners of the whaler and cutter races. For weeks crews had been in serious training, up at the crack of dawn to row up and down Grand Harbour or Sliema Creek. Many hours had been spent sandpapering boats and oars to remove the excess weight. All this preparation would now come to fruition! The bookmakers set up their black-boards, and a carnival atmosphere prevailed for several days. Banyan parties were held on the beaches, with beer brought in by Tank Landing Craft. At night raiding parties from the losing ships would try to board the winners but would be repelled in most cases by well directed hoses or potatoes.

In Group One, HMS *Phoebe* came first, HMS *Forth* second,

HMS *Gambia* third, and *Glory* fourth. It was a very relaxing and enjoyable time. Cyprus was the next island to be visited, and *Glory* in company with the frigate HMS *Pelican* anchored off the small seaport of Larnaca on Friday 14 July. In the morning the Commissioner, Mr Aldridge, and the Mayor, Mr Lyssos Santamas, called onboard to see Rear-Admiral Grantham, who returned their visit later in the day. The Commissioner and Mrs Aldridge gave a dance that evening in honour of the Rear-Admiral and officers of the two ships. On Saturday night the canvas awnings were rigged and a dinner dance was given onboard *Glory*. There was an incident when one rating was brought back from shore suffering a stab wound.

The following Saturday *Glory*, in company with HM Submarines *Token* and *Sturdy*, the frigates *Peacock* and *Loch Dunvegan*, the Royal Fleet Auxiliary *Fort Duquesne*, and Submarine Depot Ship *Forth*, received a great welcome when they arrived in Alexandria. After the usual salutes Rear-Admiral Grantham went to Raz el Tin to sign the official visitors' book, and then called on Admiral Badr Bey, the Commander-in-Chief, Royal Egyptian Navy, and later on His Excellency the Ambassador, Sir Ralph Stevenson. Formalities over the social activities got under way with a cricket match, and over the following days there was a water-polo match against the Alexandria Sporting Club, and a regatta at the Hellenic Rowing Club. The New Sports Club held a cocktail party at which nearly 300 members and guests were present.

On the following Tuesday the Egyptian Press were invited to visit *Glory* and the other ships. When HMS *Glory* arrived back in Malta on Wednesday 2 August, the splendid looking Italian battleship *Doria* was in Grand Harbour. The big event during August was the twenty-one gun salute fired by *Glory*, *Cheviot*, *Chieftain*, *St. Kitts*, *Cadiz*, *Sluys*, and *Forth* on the occasion of the birth of Princess Anne, on Tuesday 15th. In September, the ship paid a visit to the French seaport of Marseilles. *Glory* unfortunately was given a berth on the mole at a considerable distance from the city, whereas the destroyer escort were allocated berths almost on the main thoroughfare. Marseilles with its surrounding hills and offshore islands was a fascinating city to explore.

The old university and cathedral were of particular interest and many visited the old quarter, much of which had been destroyed by the Germans in 1943. The splendid shops, bars and clubs along La Canbiere were also well frequented. The visit ended on Wednesday 13 September, and after three days of exercises with American, French and British ships, *Glory* arrived off Tangier on Saturday in company with HMS *Saintes*. Tangier, abandoned by the British in in 1684, became a centre for Barbary pirates among whom were a few well bred Englishmen. The narrow alleyways of the Kasbah, the Souk, and the strong Spanish influence made it a very intriguing place for a run ashore. As the ship was anchored some distance from the jetty a Motor Fishing Vessel from Gibraltar was attached for liberty trips. The final port of call was once more at Gibraltar.

Lord Hall, the First Sea Lord, inspected the ship's company on Monday 25 September, after which there was a march past on the flight deck. During the visit football, hockey, water polo and rugby matches between ships and fleets took place. The stay in Gibraltar marked the end of the Mediterranean and Home Fleet exercises, and ships began to disperse during the week. HMS *Glory* returned again to Malta. It had been a very busy year, much flying had been accomplished, and many thousands of miles steamed. Most of October was spent in the floating dry dock in Grand Harbour for a general overhaul, after which came the ritual of paint ship, with music while you work performed by *Glory*'s Royal Marines Band.

A large number of *Glory*'s ship's company at this time were National Servicemen, and for most of them 1950 had been a very enjoyable experience. Not for them the dreary days spent marching up and down some windswept parade-ground waiting for demob. The sunshine, the excitement and danger of naval aviation, the visits to exotic ports, would be memories that would stay with them all their lives and in fact would seduce some of them into 'signing on'. For weeks 'buzzes' had been doing the rounds that the ship would be going to the Far East in 1951, to relieve HMS *Theseus*. The 'buzzes' were eventually confirmed, and in November much activity took place with tons of stores and many vehicles being loaded aboard for passage to the United Kingdom. The 14th Carrier Air Group would remain at Hal-Far until the ship returned in February 1951. At last *Glory* with her long paying off pennant flying left Grand Harbour with a few hundred well wishers waving from shore.

The voyage home was horrendous. On Wednesday 13 December the ship ran into very heavy seas and for the next twenty four hours was severely battered by savage gales in the Bay of Biscay. Waves were shooting high above the bows and breaking over the flight deck amidships, the force of cascading water smashing boats and breaking a bulkhead door. Messdecks were swamped and a flight-deck gun sponson was crushed. Stoker Mechanic G.J. Causon, of Cransham, Gloucestershire, suffered a fractured skull during the rough weather which forced *Glory* to heave-to for twelve hours. Several other ratings suffered minor injuries.

It was a forlorn looking carrier which eventually berthed in Devenport Dockyard on Saturday morning 16 December, her paying off pennant streaming in an icy wind which chilled the waiting crowd of relatives, and the Royal Marines Band played the highly appropriate strains of 'Never mind the weather, here we are again'. The ship tied up above HMS *Ocean* over twenty four hours behind schedule, which led to many wasted journeys by some relatives and a long wait for others. Mr and Mrs Soper made their second trip from Paignton in two days to meet their son Leading Air Mechanic Alan Soper, who had returned after being stationed in Malta, for his first Christmas at home for four years. Lieut-Cmdr J.A.J. Smith-Shand was greeted by his wife and small daughter who had been waiting in a Plymouth hotel since Wednesday, having travelled down from London early in case *Glory* berthed ahead of time.

Mr H.S. Spurgeon was on the quayside to welcome his son Lieutenant P.L. Spurgeon Royal Marines, having spent a day longer in Plymouth than he expected. Some of the ship's company had completed a two-and-a-half year commission which they began in HMS *Ocean*, transferring to *Glory* in November 1949. By the following Tuesday *Glory* lay silent, battered and rust streaked after her storm-tossed crossing of the Bay. The old ship's company had dispersed for Christmas leave, and just a care and maintenance party were left aboard. The damaged sponsons were repaired and the smashed boats replaced before the New Year.

7

Korea & Australia
(January 1951 – January 1952)

HMS *Glory* recommissioned at the end of December 1950, with a special war complement for operational duties with the United Nations Forces in the Far East. It was at midnight on 29 December 1950 when the special train left Chatham Dockyard, carrying the main body of *Glory*'s ship's company to Plymouth, where they arrived the following morning in steadily pouring rain. They embarked straight away, and settled into their parts of ship. Throughout the next few weeks others would join in dribs and drabs. The ship was now under the command of Captain Kenneth Colquhoun DSO RN, he was over six feet tall and his purposeful facade housed a quietly competent personality. He had joined the Royal Navy as a cadet in 1918, and served in the *Ark Royal* when she commissioned in 1938 until after the Norwegian Campaign in 1940. He commanded the Escort Carrier HMS *Trumpeter* on Atlantic and Russian convoys, and in a strike against the *Tirpitz* in northern Norway in August 1944.

Captain Colquhoun learned to fly after the war and had his first solo flight on his 44th birthday in 1948. He commanded the Royal Naval Air Station at Culdrose, in Cornwall, before becoming Captain of *Glory*. In practically everything from purpose to power, a submarine and aircraft carrier are just about at the extremes of naval opposites, of which Commander Robert Alexander RN, was no doubt acutely aware when he first stepped into *Glory*'s vast innards as second-in-command. Commander Alexander had spent most of his service life in submarines, and towards the end of the war was the first Captain of the ill-fated *Truculent*.

When *Glory* arrived in Grand Harbour on Friday 2 February, the jovial reign of that merry monarch, King Carnival, was in full swing. The inhabitants flocked from Sliema, Hamrun and all the other villages in the island. The Maltese are past masters of the arts of papiermache masks and heads, and grotesque figures pranced and danced. Floats filled with girls in glorious costumes, music, whistles, streamers, flowers and confetti. The carnival was an opportunity for all the local firms to advertise their goods, and 'Simons Farsk' the brewers had a huge Neptune holding an equally immense anchor, advertising their brand of the matelots' favourite beer 'Anchor'. Vast crowds came into Valletta to watch the parades and dancing on Palace Square, and the carefree revelry continued until Tuesday when on the stroke of midnight, the busy whirl of make-believe came to a close.

The *Glory*'s very busy working-up period began with flight deck drills, anti-aircraft gunnery, damage control exercises and many other evolutions. The two Squadrons which made up the 14th Carrier Air Group were also extremely active. The Sea Fury aircraft of 804 Squadron were commanded by Lieut-Cmdr J.S. 'Bill' Bailey RN, and 812 Squadron were equipped with Firefly aircraft and commanded by Lieut-Cmdr F.W. Swanton RN. Despite some very rough seas on one day during the week, bombing runs, rocket and gunnery attacks were constantly carried out. Three aircraft came to grief whilst landing but there were no casualties and the damaged planes were put ashore by lighter when the ship anchored in Marsaloxx.

The ship spent all week at sea returning to either Grand Harbour or Marsaloxx at weekends. There were the occasional diversions from the unremitting noise of aircraft taking-off or landing, and on 12 February, the *Glory* anchored off St Paul's Bay to fire a twenty-one gun salute in honour of Princess Elizabeth who was returning to England. There was great joy in 18 Mess (Flight Deck Party), when the mess cat gave birth. The kittens looked very similar to some of the rats sighted around the ship, but all the maternal instincts of the mess were aroused and they tried to tempt the mother onto a sack padded with cotton waste but she preferred the dusty deck beneath the lockers. At the end of February, *Glory* was anchored in Marsaloxx with the USS *Franklin D. Roosevelt*, a huge aircraft carrier equipped with 5in guns, 'Banshee' jet fighters, twin engined 'Neptune' bombers and the inevitable helicopter. *Glory*, after two very rough passages across the Bay looked rusty and shabby in comparison with the American carrier's pristine condition.

The first few days of March, however, saw the ship in Grand Harbour with all hands employed painting ship, dressed in anything from shorts and a bowler hat, to fezzes and sea boots. Paint-covered apparitions wandered around carrying 'long-Tom' brushes, and eventually the ship emerged resplendent in new coat of light grey. The frenzy of preparation for war continued, and 'Operation Alert' was carried out with the port watch acting as look-outs during a mock attack by the 'enemy' frogmen.

The 14th Carrier Airgroup and their 21 Sea Furies and 15 Firefly aircraft were embarked and *Glory* was almost ready to assume her wartime duties. At 0001 precisely, on Tuesday 20 March 1951, the ship slipped away into the darkness en route to the Far East. At Port Said, gateway to the eastern world, hordes of 'bum boats' thronged around the ship and although trafficking was illegal much surreptitious bargaining was conducted through the for'd heads scuttles. Much of the Canal was traversed during daylight and proved very interesting for the majority of the ship's company who had never experienced the familiar banter of Army and Air Force personnel and their witty quips as the ship passed slowly by, 'Get yer knees brown jack', and 'When is the *Mayflower* coming through?' Out into the Red Sea which was like a sheet of glass. It was extremely hot and lime juice, salt tablets and iced water, when available, helped to make conditions a tiny bit less intolerable.

There were two small canvas pools measuring about 7ft by 7ft rigged on the cable deck, and somehow they accommodated the entire lower deck! Hordes of naked bodies jostled each other in the murky brown water. Flying continued from first light to 1800, and everyone who worked on the flight deck began to look lean and brown, whilst those employed below looked dreadfully pallid when they appeared in the bright sunshine. *Glory* passed through the 'Gates of Hell' and headed towards Aden, where there would be a very short stay to take on fuel oil and replenish the water tanks.

Aden, swarming with humanity from practically every nation on earth, with its small passageways, dark and evil smelling, and its population dressed as though from the contents of the scran-bag. The cleanest building in Aden was probably the United Services Club which had a small swimming pool and a large air-cooled terrace and restaurant, and also had a section of sea fenced off from sharks. Most who went ashore enjoyed a swim and change of scenery, dull though it was. Back at sea on March 30, and the 1,000th deck landing was completed. A Firefly coming into land later in the day was unable to lower either flaps or wheels, despite carrying out some desperate manoeuvres all over the sky to try and dislodge them. Under the circumstances the landing was very decent indeed, with just a small outbreak of fire which was quickly extinguished.

Onward across the Indian Ocean with shoals of flying fish skimming along the surface for up to 30 yards or so. A Firefly catapulted into the sea on Tuesday, but fortunately both the crew were picked up by the destroyer HMS *Gravelines*. A picturesque view of Malaya was seen as *Glory* steamed through the Straits of Malacca, small grass huts, palm trees and superb beaches with the dramatic back-cloth of deep green jungle. There was a short stop in Singapore to take on stores and aircraft, then the ship headed north for Hong Kong on a sea with hardly a ripple, where there were more flying fish and sea snakes in abundance. Nine days were spent in Hong Kong completing the transition from a peacetime ship to a wartime one.

Everyone had time to discover the delights of the Colony. Shops galore, journeys up the Peak Railway gave a beautiful view of the harbour below with ships laid out like toys, the Star Ferries to Kowloon, and of course, the China Fleet Club with its 'big eats for the boys'. Entertainment was also provided onboard by Chinese acts, one of which was a family of acrobats who performed with amazing dexterity in the forward lift-well; at one time Grandma aged 62, supported the five other members of the group. On 20 April *Glory* left Hong Kong and headed for Japan, the inevitable flying took place with the extra sideshow of the dolphins sporting themselves alongside, leaping clear of the water, their silvery backs shining in the sunshine.

When *Glory* arrived in Sasebo harbour on Monday 23 April, the aircraft carriers HMS *Theseus* and HMS *Unicorn* were anchored ahead and handing over routines began immediately, with aircraft being brought over from *Theseus*. Almost all the ship's company made their way to the flight deck to inspect *Glory*'s latest acquisition, a Sikorsky helicopter, kindly loaned by the United States Navy. It was crewed by Lieut P. O'Mara USN, and CPO Fridley USN. 'The Thing', as it was christened, carried US Navy markings along the fuselage but in addition had *Glory*'s crest painted by the cabin, and would prove to be a valuable asset to the ship. HMS *Theseus* left Sasebo on Wednesday, *Glory*'s ship's company lined the flight deck to cheer her on her way home, and the band on *Theseus* played what had become our unofficial signature tune, 'Glory, Glory, Hallelujah'.

Sasebo, situated in the north west of Kyushu Island, Japan, was quite a picturesque seaport surrounded by green hills and would be used as a base during the ship's operational duties for the United Nations. HMS *Glory*'s main task during the war was to operate off the west coast of Korea where her aircraft could attack enemy supply lines and troop concentrations. The ship would leave her base in Japan, escorted by Canadian, Australian, New Zealand, British or United States destroyers, and arrive in the operational area a day later. Weather permitting, sorties would then be flown from first light until dusk during the succeeding four days. Re-fuelling usually took place on the fifth day from a Royal Fleet Auxiliary tanker such as *Wave Premier* or *Brown Ranger*. *Glory* then resumed her duties for a further four days, then handed over operations to the American aircraft carrier with which she alternated.

Operations off Korea were followed by the voyage back to Japan, making eleven days at sea. If circumstances permitted the ship spent five or seven days in harbour which were almost as hectic as those at sea, as damaged aircraft had to be transferred to HMS *Unicorn* and replacements obtained; ammunition and provisions had to be embarked, and necessary maintenance work undertaken. There were occasions of course when the programme needed to be revised at very short notice, and time spent in harbour could be reduced in extreme circumstances to only two days, calling for extremely good team work from all the ship's company.

HMS *Glory* left Sasebo on Thursday 26 April to begin her first patrol. Fog and rain prevented any flying on the first day, but on Saturday the hands were piped to flying stations at 0430hrs and the long operation began. During the day a Sea Fury piloted by

59

Lieutenant Edward Stevenson RN disappeared, and after a long and diligent search by escorting ships and the helicopter, he was assumed to have crashed into the sea; it was a sombre beginning. On Monday enemy forces occupying the Kimpo Peninsular at the western end of the front were shelled by the cruisers *Belfast* and *Toledo*, and strafed by *Glory*'s aircraft.

Tuesday was spent re-fuelling, but on Wednesday Corsairs and Skyraiders from the United States carrier *Bataan* joined the Firefly and Seafury aircraft from *Glory* to attack more targets. During this operation Lieutenant Barlow had to force-land his Sea Fury in the River Han. Fortunately he was not injured and was quickly rescued by helicopter. Every two hours throughout the day a sortie would take off and the previous one land on, and within a very short time 50-60 sorties a day were being carried out without undue strain on the ship's handling and servicing capabilities. Another feature of this first patrol was the considerable amount of flak which the aircrew experienced, from 88mm guns to massed rifle fire, and they rapidly learned not to hang around after pressing home their attacks. On Sunday 6 May, after carrying out more attacks on enemy positions HMS *Glory* started the voyage back to Sasebo, much wiser from the experience.

Incredibly there would be only two days in harbour! Soon after arrival there was frantic activity as replenishment of ammunition, stores and victualling began. One of the better duties in harbour was to be detailed as 'Pier Patrol', which was certainly exciting. The patrol had the job of trying to contain a crowd of Australian, Canadian, 'Kiwis', and British matelots, who were behind two barriers with a gangway between. The idea was that several hundred libertymen who were invariably well under the influence of drink, would stand in orderly fashion until their respective ship's boats arrived to take them off. What happened on each occasion was a shouting, singing, fighting, hollering horde, who smashed down the barriers in their efforts to get aboard their boats!

HMS *Glory* left Sasebo on Thursday 10 May, and was straight into the fray again the following day when the Firefly and Sea Fury aircraft attacked bridges and villages. It was during this second patrol that cartoons started to appear in the canteen flat showing

Glory's aircraft strafing ox-carts with rockets and cannon shell. Such a method of transport was, however, a very effective way of carrying stores and ammunition, and many exploded with tremendous force after being hit.

The RFA tanker *Wave Premier* re-fuelled the ship on Monday 14th, and whilst doing so Able Seaman J. McPherson of Strood, Kent, fell overboard from the quarterdeck. Luckily the ship's photographer, Petty Officer E.J. King, of Pinner, Middlesex, was airborne in the helicopter and noticed a commotion and spotted the life-belts floating astern. The helicopter came down over the spot where McPherson was struggling in the cold water, and Petty Officer King, wearing his Mae West, dived to the rescue and managed to support the half drowned McPherson until a sling was lowered and the helicopter landed him back aboard *Glory*. PO King was picked up by a Canadian destroyer. Both men suffered slightly from their exposure in the water but later returned to duty. A very welcome sight during operational patrols was the arrival alongside of a destroyer carrying mail. The bags were transferred by jackstay and this evolution was watched anxiously by members of the ship's company as sometimes the bags came perilously near to a soaking.

A most welcome piece of news was received from the Admiralty, who stated that it had been found desirable to give ships' companies and aircrews on aircraft carriers a period of rest after six months' service in Korean waters. If the war in Korea continued, said the statement, HMS *Glory* would require some relief in October, and to effect this, while avoiding even a temporary reduction in the British Commonwealth contribution to the United Nations forces in Korea, the Australian Government had generously agreed to make HMAS *Sydney* available for Korean service for three months from October. What most people wanted to know was where *Glory* would spend her three months' rest! Meanwhile the air strikes continued with the ubiquitous ox-carts coming in for more punishment, and air spotters from *Glory* also assisted HMS *Kenya* to shell the area west of Songhwa, near Chinampo.

Two of *Glory*'s aircraft were lost to enemy action during the patrol. On Tuesday 15th, Lieutenant J.A. Winterbotham RN was forced to ditch but was picked up by sampan and he returned to the ship next day. On Friday a Firefly also ditched, but being less than a hundred miles from *Glory* both members of the crew were picked up by the ship's helicopter.

On arrival in Sasebo it was necessary for the ship to be dry-docked so that repairs could be carried out to the packing glands on the propeller shafts. There was a great deal of inconvenience caused when a ship with over a thousand men aboard went into dry-dock; in particular, the ship's heads were locked and everyone had to proceed ashore when the need arose. During the day this caused no great difficulty, indeed it made a welcome break for inquisitive matelots to have a mooch around the dockyard at the same time. But during the silent hours people were very reluctant to stray far from a comfortable hammock or bunk, and so they usually tried to find a nice quiet place from which to relieve themselves into the dock, which enraged the dockyard workers when they arrived for work! To alleviate this problem it was the responsibility of the duty watch to place very large buckets at strategic points on all weather decks throughout the ship at pipe-down. In the morning it was the men under punishment who emptied the buckets ashore. It was a very delicate operation, fraught with danger, for two men carrying a bucket full to over-flowing down a steep and slippery gangway. Fortunately the ship only spent three days in dock.

Japan is a beautiful country, and during *Glory*'s extended stay in Sasebo many were able to get ashore and explore the place. Climbing to the top of one of the hills which overlooked the harbour ensured a breathtaking view. In the distance the blue sparkling sea seemed to push into the land like fingers in wax, and the myriad small islands, lush with greenery lay directly below like folds in a quilt; it was a glorious panorama.

In the town tiny school children on their way home would stare in fascination at the tattoos on sailors' arms, and would follow behind holding their eyes open in imitation of the western rounder eyes. They were also intrigued by the bridges of matelots' noses as they felt their own much flatter profile. In the dockyard, Japanese men wore peculiar footwear, with a division for the big toe, rather like a shoe fashioned in the shape of a mitten. Those wearing trousers covered the leg from ankle to knee in puttees reminiscent of first world war soldiers, while those in shorts wore socks with suspenders! A very pleasant custom with Japanese people is bowing, which did not seem servile in any way but more of courtesy in old world style.

Bus trips were organised to Nagasaki, the seaport about a hundred miles from Sasebo which was destroyed by atom bomb on 9 August 1945. The roads were in a dreadful state and extremely bumpy to ride over, but the countryside was well worth seeing with its patchwork of little fields, every inch of which was cultivated by the industrious farmers. The atom bomb site was merely a board proclaiming the centre of the holocaust, with a small museum containing grisly relics.

After fourteen days' recuperation in Sasebo, *Glory* departed once again for the operational zone on Sunday 3 June, and the following day everyone slipped easily into familiar routines: the early morning hustle and bustle of preparing aircraft for the first detail in the dismal darkness before dawn; the tremendous racket of the catapult, and revving engines as each aircraft was launched; the eerie silence when the last aeroplane disappeared over a murky horizon. Then all the hands raced below to grab breakfast and a smoke before ranging the next sortie. All went well until the last land-on. One Sea Fury was missing, and later it was learned that Lieutenant P.A. Watson RN had ditched and been picked up by the frigate HMS *Black Swan*. On Tuesday, Pilot 3 Stanley Ford returned from a sortie, and as he prepared to land his engine developed a fault and he ditched not far from the *Glory*, the aircraft sank and Mr Ford was not able to escape.

A Sea Fury landed and taxied up the deck, but could not stop and ripped half the tail plane from another aircraft. It was discovered that the Sea Fury had been hit by bullets in the port wheel, which affected the braking system. The *Wave Premier* re-fuelled the ship on Saturday and apparently transferred aviation spirit which was contaminated. This incident caused 'buzzes' to circulate about sabotage, but it later transpired that the cause was a corroded pipe, not enemy agents. There was another incident at the end of the third patrol when an aircraft made a very erratic approach and was badly damaged after landing heavily, but no one was injured.

HMS *Glory* returned to Japan but this time to the naval base at Kure, in the south west of Honshu Island. The journey through the Inland Sea provided the ship's company with views of 'true' Japan, the scenery was exquisite with even the tiniest of islands having a tree perched crazily on it. The ship arrived on Friday 15 June and secured to a jetty directly opposite HMS *Unicorn*, which towered above *Glory*. The ship's company found Kure an infinitely better place than Sasebo, as it was mostly occupied by Commonwealth troops and the facilities did not have the gaudy brashness which seemed to accompany so many American bases. Many were able to sample the delights of a steamboat trip to the small and very peaceful island of Miyajima, with its excellent beaches, and where woodcarving with the most intricate workmanship could be purchased for a few hundred yen, or the most terrifying carved skulls with a snake twisted through its hollow orbits could be purchased.

The ship was replenished with ammunition and stores and left to start the fourth patrol on Thursday 21 June. Admiral Scott-Moncrieff came aboard the next day to observe for himself the efficiency of *Glory*'s operations. The following Thursday the ship's company were saddened to learn that Lieutenant John Sharp RN and Aircrewman George Wells were both killed after their aircraft was shot down. Three more aircrew were lost during the fifth patrol: on 16 July, Lieutenant Robert Williams RN and Sub-Lieut Ian Shepley RN died when their aircraft was shot down, and two days later Commissioned Pilot Terence Sparke RN was killed in Korea.

It was during July that Lieutenant R.H. Kilburn RN, while acting as flight leader sighted a MIG fighter in the water north of Chinnampo. For months United Nations commanders in Korea had tried by various devices and ruses to lay hands on one of the Russian built fighters, and for months the Reds had foiled them. The position of the aircraft was fixed and photographed and a marker buoy dropped by helicopter. The frigate HMS *Cardigan Bay*, a South Korean motor launch and a US shallow draught landing craft equipped with a crane moved in through treacherous sand bars to retrieve the prize, while a cruiser and *Glory*'s aircraft stood by to ward off enemy interference. Darkness and high tide interrupted the operations and the allied craft had to stay on the spot, but next morning they got the MIG onboard and made off with it safely.

The MIG was sent to the US experimental base in Dayton, where it was discovered that the engine, which everyone had feared was a redoubtable fruit of Russian plus German technology, was an unmodified British-made Rolls Royce Nene! Britain had shipped perhaps 100 jet engines to Russia before trade was stopped in 1948. The dawn to dusk flying continued with only the occasional interruption caused by defective equipment. *Glory* had only the one catapult designed originally to launch aircraft of 14,000lbs at 66 knots. Under Korean operational conditions a Sea Fury carrying out air reconnaissance weighed 13,200lbs, and one equipped for attack 14,200lbs. A Firefly under most conditions weighed 16,000lbs.

These extra requirements created a heavy strain on the machinery and often caused breakdowns due to pipe fractures and leaking joints. During two years of operations, *Glory* carried out 8,302 heavy launches, while at the receiving end 10,012 safe landings were made. These achievements were made possible by the spirit and team work of the flight deck personnel of the Engine Room Division, who worked throughout the day whenever there was flying. They worked in one continuous watch and were frequently 'on deck' for fifteen hours. Many hours of 'overtime' by all became the rule rather than the exception at sea and in harbour. HMS *Glory* finished the fifth patrol on Saturday 21 July and headed for Kure and what was hoped to be five days' rest and recreation.

However peace talks between the warring factions had supposedly reached a crucial level, and so only two days were spent at Kure, then it was off to the west coast again to operate with the USS *Sicily* to back up other United Nations units for a spot of 'sabre rattling' in the forlorn hope that someone would be intimidated. The two aircraft carriers operated until Sunday, when *Sicily* departed for Sasebo. There was some very peculiar weather during the patrol, fogs, sea mists, and torrential rain sweeping across the sea in dense sheets, with the ship emerging seemingly rustier every time. Flying was of course severely restricted but even so some targets were attacked with good results. The USS *Sicily* returned to the area on Saturday 4 August, and *Glory* set off once more for the dubious pleasures of Sasebo.

The seventh operational patrol was another which suffered badly from adverse weather. *Glory* managed to get five days in Sasebo, leaving on Friday 10 August. The ship made her way into the Inland Sea and picked up replacement aircraft from the airbase at Iwakuni, continuing then to Kure for a one day stopover before steaming again for the west coast. On Thursday 16 August flying was brought to a close because of the deteriorating weather, and as the day wore on conditions onboard began to get very uncomfortable. The following day Admiral Scott-Moncrieff came aboard and *Glory* then gathered in the rest of the fleet and retreated across the Yellow Sea to the Shanghai Approaches, thence down to Okinawa, where to everyone's relief the ships finally evaded the typhoon's attentions.

According to reports, *Glory* was never nearer than 150 miles to this trick of nature, but that was quite sufficient. For two days and nights the mess-decks were flooded, weather decks covered in oil and water, and at times the angle of the flight deck was at 45 degrees; the stench below was unbearable, and life became very miserable. But anchored safely in Buckner Bay, Okinawa, on Tuesday 21 August, the sun shone brightly and the hands were piped to bathe over the port side for'd, soon everything was back to normal. The following day the ship was rolling back to Kure in a huge swell with the flight deck scorching under the blazing sun. The things that made life bearable on *Glory* were very simple, an occasional trip ashore, a 'make-and-mend', and mail from home.

Back in the operational area again on Monday 2 September, flying resumed with great intensity. On Sunday, an all out flying effort was made which resulted in *Glory* setting a record of eighty-four sorties in one day. It was a day which will always be remembered by Lieutenant Percy Morris RN who, with his wingman, was flying their aircraft as an extra section to make the required total on the last strike of the day. He said, 'Our target was a small but fairly important railway bridge. It had been attacked unsuccessfully earlier that day. No anti-aircraft opposition had been noted. It was therefore decided that we would drop our two bombs simply making two runs each. However at the end of my first steep attacking dive there was a loud bang and my windscreen was covered in oil. My bomb landed I know not where as I used my speed to get clear and gain height. At some 7,000 feet I had slowed down enough to open my cockpit hood and gain some visibility. My engine had no power so I headed for the coast not far away, aiming to go as far as I could. My Observer and I checked off the vital actions – jettison drop tanks, switch off fuel, cut the magnetos etc. We soon reached an area of rolling sand dunes very near the sea and I made what I claim to be a rather smooth landing, wheels up of course, and we skated to a gentle stop. We both leaped out, taking only our personal weapons and dinghies. Paddling out to sea would be better than hanging about on an unfriendly shore. We did not burn the aircraft as our arrival in what seemed to be an uninhabited area might not have been noticed.

'As we started off across the sand dunes my Observer pointed to an object lying in the aircraft skidmarks. "What is it?" he asked. "Must be a drop tank," I replied. While we marched along we were covered by other members of the sortie. Later on there were some Sea Furies, then some United States aircraft, Shooting Stars and Mustangs. Then to our delight there arrived a seaplane. All we had to do was paddle out and hitch a lift. However, before we could do that *Glory*'s helicopter arrived. It took very little time for the helicopter to land and for us to climb in. Once aboard we both had the same thought – have we got enough fuel to reach *Glory*? "Yes" was the reply, "we've got enough." Despite this assurance, however, after a very short while the needle gave a sudden flick and dropped to zero!

'We were very near a little island called Piapdo so the pilot landed on a level area clear of habitation. We wondered if the gauge was correct so we lowered a penknife into the tank. All we heard was a hollow clank. Very soon a group of local gentry with traditional tall hats came cautiously along with hands raised. The ladies stayed at a safe distance. But, we had no cigarettes and they had no petrol. We had noticed a motor sampan heading off to the mainland, no doubt to report our arrival. Our fears were soon dispelled by the appearance over the horizon of a destroyer. It was HMCS *Sioux* to the rescue. The ship landed an armed party to guard the helicopter and took us back to *Glory* to collect aviation petrol.

'Next day, at first light, the helicopter returned safely to *Glory*. So ended an eventful few hours. I took my time before I told my Observer that the "drop tank" on the dunes was a 500lb, bomb that I had failed to jettison! It's easy to look back on it with a laugh, but then I remember that in six months of operations we lost five killed and one wounded in our squadron alone and laughter seems inappropriate.'

With the helicopter back aboard, *Glory* returned to Kure with everyone feeling very elated by the record achievement. It had been confirmed that the ship would be going to Australia for a well earned rest sometime in October, which added to the rather joyful atmosphere throughout the ship.

There were many people though who had been serving in *Glory* since the middle of 1949, and they were due to be relieved in Singapore. In anticipation of that event they had been busy painting black anchors and 2in wide red lines around their kitbags and hammocks – this was a requirement for Royal Navy personnel travelling home by troopship – with great care. *Glory* left Kure on Sunday 16 September for her ninth patrol as part of Task Force 95, to take part in the first co-ordinated visit to the eastern seaboard of Korea. The strike was commanded by Rear Admiral George C. Dyer USN, aboard the cruiser USS *Toledo*. Other escorting destroyers were USS *Parks*, USS *John R. Craig*, USS *Orlech* and USS *Samuel N. Moore*. The strike commenced on Tuesday, with an attack on Wonsan.

The next day the catapult became unserviceable which meant that the dreaded Rocket Assisted Take Off Gear (RATOG) was brought into use. It could be argued that this equipment caused more anxiety to aircrew than the flak over Korea. A group of rockets were strapped each side of the fuselage close to the wing roots. The aircraft would then begin a 'normal' take-off run during which the pilot would fire the rockets which would 'assist' the aeroplane off the flight deck in spectacular fashion. It was noisy, smelly, and extremely dangerous if the rockets misfired leaving the pilot fighting to fly in the right direction as well as gain altitude. It was whilst using RATOG on Saturday 22 September, that a Firefly aircraft with a 500lb bomb under each wing plunged over the ship's bows when the rockets failed to ignite.

The pilot by a miracle survived but his Observer, Sub-Lieutenant Ronald Davey RN, was drowned. It was an appalling thing to have happened on the last patrol. The following Tuesday was the last day of operations and the last aircraft to land on was piloted by Lieutenant Commander Swanton, who was given a rousing reception as he stepped from his Firefly. *Glory* arrived back in Kure on Thursday to find that her relief, HMAS *Sydney*, had arrived and was lying opposite. The ship's company set-to with alacrity to transfer stores and aircraft across to *Sydney*. Then there were last minute runs ashore to purchase a Japanese tea set for mum, and finally divisions and inspection of the ship's company by the C-in-C Far East Station, Vice-Admiral Sir Guy Russell.

During those last days in Kure it seemed as if a halo of euphoria surrounded *Glory*. All the tensions, the trials, the tragedies, of the last six months were being put aside as most looked forward to Australia, whilst the remainder eagerly made their preparations for going home. Naval Airman Anthony Hammond was especially pleased. He had just celebrated his coming of age, which was the third birthday he had celebrated onboard *Glory*. He had joined the ship in Devonport on his nineteenth birthday just before it left for the Mediterranean; his twentieth had been spent at sea between Gibraltar and Malta, and now his twenty-first, in Kure. At tot time his mess-mates were generous with their sippers and gulpers, and after dinner Anthony was laid to rest under the mess table where he slept until reveille next day.

At 1630 on Sunday 30 September 1951, HMS *Glory* left Kure and headed for the fleshpots of Hong Kong. Lower deck was cleared for entering harbour on Wednesday 3 October, but as the ship prepared to secure it was announced that Engine Room Artificer D.E. Dixon had been lost overboard the previous evening; it was a sombre arrival. The harbour was bustling as usual, sampans bobbing around the ships at anchor, and junks with their prominent stems and lug sails heading for a spot of trading or smuggling. There were freighters, tramp steamers, ocean liners, and the ferries continually puttering to and fro with their packed cargo of curious passengers, and naval craft speeding between the different warships.

Hong Kong harbour at night was a splendid sight with everything lit up. The houses on the tops of the surrounding hills looked for all the world as if they were suspended in mid-air. There was time to purchase more 'rabbits', dressing gowns dripping with Chinese Dragons, or little wooden rickshaws. After two days the *Glory* left for Singapore. On Sunday at sea, after 'divisions' in the morning, a Sunderland flying boat carried out bombing runs on a towed target astern. There was a frenzy of activity in Singapore on Tuesday, as personnel completing their commissions left for the Royal Naval Barracks to await passage to the United Kingdom, whilst their reliefs settled in on *Glory*. It was decided to hold a 'Crossing the Line' ceremony soon after leaving Singapore, but foul weather marred the proceedings.

King Neptune and his Court were at first delayed by rain, and although *Glory* was only a few miles from the equator it turned so cold that the 'Initiation' had to be abandoned. There were other delights to be had! The 14th Carrier Air Group staged several performances of the Revue which had been written and rehearsed entirely in their own time, and which proved a great success. The ship made a brief call at Fremantle, Western Australia, on Wednesday 17 October, for fuel, water and that most popular of commodities mail. As *Glory* continued steaming along the south coast of Australia, the Air Group was busy packing its bags in preparation to fly off to HMAS *Albatross*, the Australian Naval Air Station at Nowra, 100 miles south of Sydney.

Gale force winds delayed the unloading of stores and personnel of the Air Group when *Glory* arrived in Jervis Bay, on Tuesday 23 October. The gale, which blew at a velocity of 50 knots at times whipped up the waters of the Bay, and it was impossible for naval lighters and workboats to come alongside. At about 1800hrs, the winds abated and work proceeded at great speed to land 280 officers and men of the Air Group plus baggage and three unserviceable aircraft. The ship arrived in Sydney the following day. Captain Colquhoun had already agreed that HMS *Glory* should take part in the Australian celebration of Trafalgar Day which was being used as a recruiting drive for the Royal Australian Navy, and the anniversary was postponed until Saturday 27 October, to fit in with the ship's movements.

Despite the high tempo of preparations for the refit a creditable show of live and static displays was given onboard, and some 15,000 people who swarmed over the ship were presented with free souvenir programmes. It was a day which saw the beginnings of many lasting friendships, and some half a dozen ratings got married and many volunteered for loan service with the Royal Australian Navy. The full measure of Australian generosity was apparent when everyone wishing to proceed on 14 days' local leave was offered the choice of more than one address at which to stay. Bus and tram travel in the city was free and a generous allocation of free tickets for cinema and attendance at Test and other cricket matches was made. So began a period of relaxation.

8

Korea & Mediterranean
(May 1952–October 1952)

The word relaxation seems rather mild in view of the boundless hospitality that was lavished on the ship's company. The fourteen days' leave was spent in numerous ways, perhaps the most popular being a stay on remote stations way out in the bush, an unforgettable experience. The ship was taken in hand for refit on Monday 29 October, and was in Captain Cook dry dock between Monday 5 November and 6 December. During the refit the two midshipmen's chest flats were converted into five berth dormitories complete with washbasins, and ventilation to the starboard dormitory was improved by the provision of an additional fan and trunking. A new meterological office was built in the starboard side gallery deck immediately for'd of the TB Ready Room.

The stern link bushes were withdrawn for re-wooding and both stern glands re-packed, which gave less vibration at speed. The starboard HP Turbine top half was lifted to make good a leaking joint, and examination of the turbine revealed a satisfactory condition. All boilers were re-bricked throughout and cleaned internally. All catapult machinery was thoroughly overhauled, but during deadload trials it was observed that the towing bridle still tended to slacken very slightly during the early stages of the launch. Subsequent experiments showed that a smoother and more satisfactory launch resulted if a one-eighth inch washer was inserted behind the carrot valve.

The refit was completed on Wednesday 19 December, but on the previous day in answer to public demand 500 officers and men of *Glory* marched through Sydney.

Thousands of cheering Sydney citizens lined the streets as the ship's company, led by the Royal Marines band jauntily swept by. In the presence of service chiefs and civil officials, the Honourable R.W. Street, the Chief Justice and Lord Governor of New South Wales, took the salute. The Reverend Richard Knight had also recently joined the ship to relieve the Reverend Charles Birtles, but it took some time before he got to know his new shipmates as most of their time was occupied ashore. Some were indulging in that well known pastime of 'strangling the baron', and others were so much 'up homers' that they only came onboard for duty. Most of the remainder were enticed ashore to enjoy the lovely Australian sunshine and to watch cricket or tennis, or disport themselves on the sandy beaches. The names Bondi, Coogee, Palm Beach, and Manly will stir a chord in the memory of many of *Glory*'s company.

After basin trials and machinery trials *Glory* put to sea on Christmas Eve for more testing but was back in harbour for the Christmas celebrations. Perhaps the ship's company were not really able to do full justice in repaying the many acts of kindness and generosity shown them, but a farewell party was given by the officers of the *Glory* on Friday night 28 December 1951, and was one of the gayest parties of the Christmas season in Sydney. The gangplank from Garden Island to the ship was illuminated with fairy lights, and a miniature 'Sydney Harbour Bridge' leading onto the dance floor was built over a cascading waterfall. Masses of flowers, ferns and palm trees, fringed the edge of the deck, and the bulkheads were hung with the flags of all nations. Terrifying looking 'sharks' glared out from between the flags, these were the reserve aircraft fuel tanks which had been cleverly 'redesigned' as sharks. At the supper table besides the dance floor many Chinese chefs and stewards were on duty, the main feature of the supper was enormous boars' heads decorated with flowers and Christmas and New Year Greetings. A large champagne oyster bar and many smaller bars were set up near the dance floor. The after lift had been suspended halfway between the 'ballroom' and the flight deck to form a cool mezzanine lounge where guests sat and watched the dancing, the coloured lights, and colours on the dance floor. Music was provided by the Royal Marines band, *Per Concordiam Gloria* which was recognised by many as the band which played at the White City, Sydney, on the first day of the Davis Cup Tournament.

An extremely successful dance ashore was given by the ship's company at the very fine Trocodero dance hall on Friday 4 January

1952, and an official farewell 'at home' to some 350 guests was given the same day aboard *Glory*. Finally on Monday 7 January, HMS *Glory* bid a very reluctant farewell to Sydney, everyone was sorry to leave Australia but had derived great interest and pleasure from the visit. Captain Colquhoun was distressed to discover that on the day of departure fifty-one ratings were still absent from the ship but seven deserters had been recovered before the ship left Fremantle on Thursday 17 January 1952, and another was in custody.

Nevertheless, the general behaviour of *Glory*'s liberty-men ashore was exemplary and Captain Colquhoun received the congratulations of the Flag Officer in Charge, New South Wales, on that score. The 14th Carrier Air Group had been embarked at Jervis Bay. Their stay at the Naval Air Station, Nowra, was isolated by British standards but every assistance was given in adjusting routine and arranging facilities, and the Air Group had been well cared for and did not lack for entertainment. Since leaving Japan in September twenty new pilots and three observers had joined so that the present Air Group was younger and less experienced. Heavy swell was encountered on passage between Jervis Bay and Singapore, but some flying took place with communication, interception and weapon attacks practiced.

On arrival in Singapore on Wednesday 23 January, fuel, stores (including cold weather clothing), aircraft and ammunition were embarked. The ship sailed for Hong Kong on Saturday, and for the next two days strafing exercises were carried out and two new pilots were able to get in some deck landing practice. Rendezvous was made with HMAS *Sydney* on Wednesday 30 January, when six Firefly aircraft were transferred before both ships entered Hong Kong that afternoon and the duties of the blockading force Light Fleet Carriers were turned over. Captain Colquhoun resumed the title Senior Officer 1st Aircraft Carrier Squadron, and at 0001 Zulu, on Saturday 2 February, *Glory* left Hong Kong with HMAS *Warramunga* in company.

The following Tuesday both ships arived at Sasebo, and *Glory* resumed operational duties the next day. It was as if they had not been away. The same routine was exercised as before, except that it was much colder. Railway tunnels, bridges, villages and ox-carts were attacked during the patrol and two aircraft were lost, but happily there were no casualties. After being relieved by the USS *Bairoko*, *Glory* returned to Sasebo. During the second patrol which began on Sunday 24 February, the 5,000th deck landing took place. The weather remained mostly murky with frequent snowstorms and everyone was glad when it was time to return to the more convivial atmosphere of Kure. The third patrol was marred by the death of Lieutenant Richard Overton RN, on Saturday 15 March, whilst carrying out photographic reconnaissance.

Despite being disrupted by bad weather during this period of operations, a new flying record was achieved on Monday 17th, when the magnificent total of 105 sorties was carried out in 24 hours. This was in response to HMAS *Sydney* who had beaten *Glory*'s previous record by putting up ninety-nine sorties. The ship's company were able to return to Sasebo satisfied with the result of this Herculean effort. *Glory*'s Royal Marines detachment, although small in numbers, had always made its presence felt onboard, and this commission was no exception. Their activities covered a surprisingly wide range of work and play. They had however, regretted that they were unable to land ashore in earnest, and had to sit back and watch the cruiser Royal Marines carry the torch in that respect.

Nevertheless, dawn to dusk flying operations required those who 'stand and wait' at the guns. *Glory*'s marines played their part at 'defence' and 'cruising' stations, and carried out innumerable jackstay transfers. The Royal Marines band applied itself to providing the explosives for aviators, turning with a will from bombardons to bombs, and from Irish jigs to rocket jigs. At least one 1,000lb bomb – on a rough day – was seen propelling a Band Corporal rapidly along the flight deck. Perhaps the 'Royals' most embarrassing moment came when a Marine coxswain arrived back from shore with a badly leaking boat. It had been hoisted clear of the water and a team of Royal Marines were looking for the leak outside the boat. Suddenly the corporal, who was inside the boat, shouted, 'Here it is, sir, I can see the water still coming in!'

During much of March and April, large flocks of migrating birds passed northward over the Yellow Sea, and some made perfect landings aboard *Glory* and enjoyed a brief rest. One in particular, a huge Manchurian Crane, almost identical in appearance to a stork, caused waves of consternation to sweep through the ship's company, and one Sick Berth rating was seen to examine hastily the contents of his small black bag! The bird was dubbed the 'slit eyed jumbo' and it is alleged that some confusion was caused by the pipe 'Hands fall in abreast the crane'. After a stay of two days, during which time it seemed to enjoy the general meals of cold kipper and bread laid down for it, the crane flew on to happier hunting grounds.

The next two operational patrols were dogged by storm, sea fog, thick haze, and the catapult breaking down which meant that the dreaded Rocket Assisted Take Off Gear had to be used. HMS *Glory*'s new Captain arrived to take command of the ship on Tuesday 22 April 1952. He was Captain T.A.K. Maunsell RN who had landed onboard during flying operations in a United States Navy Avenger aircraft. The next afternoon Captain Kenneth Colquhoun DSO CBE RN departed in the New Zealand frigate *Rotoiti* for Hong Kong, and thence to the United Kingdom in H.M.T. Empire Halliwell. Captain Colquhoun had been a very popular and respected man and everyone aboard *Glory* was sorry to see him leave. He now had the distinction of having commanded aircraft carriers in two wars. The ship returned to Sasebo on Wednesday 30 April 1952, and after disembarking stores and ammunition left for Hong Kong the following day having completed her second tour off the Korean coast.

The great advantage of serving aboard an aircraft carrier is that a lot of sporting activity can take place. When the *Glory* was at sea (anywhere but off the Korean coast) there was a tremendous bustle after work was over for the day. The lifts were lowered, the flight deck abreast the 'island' was cleared, and volley-ball and deck hockey began in earnest. By the end of a long passage, quite a high degree of skill had been reached in these games. A good proportion of the ship's company liked to get into the open air whether it was around the flight deck or lift wells, or on the playing pitches ashore at the different ports.

After handing over to HMS *Ocean*, HMS *Glory* left Hong Kong on Tuesday 6 May 1952, and during the voyage to Malta the transition to peacetime routine began; cleaning, painting, and polishing, for the ship had a heavy social programme ahead. Arriving in Malta on on 26 May, *Glory* said goodbye to the 14th Carrier Air Group, everyone was extremely sorry to see them go – in every way they had put on a wonderful show. Commanded by Lieut-Cmdr J.S. Hall DSC RN until December 1951, and since then by Lieut-Cmdr F.A. Swanton DSC RN, the 14th CAG had contributed greatly to the success of the allied fleet operations off the Korean coast. Their aircraft had hit at the enemy with over a million rounds of 20mm cannon shells, 14,000 rockets, and well over 3,000 bombs.

Contributing greatly to the Air Group's striking power had been the role of maintenance engineers, headed by Lieut-Cmdr I.F. Pearson RN, they had maintained almost 100% availability of aircraft throughout the two arduous tours of operations. The cost of the Air Group's effort had been twenty-seven aircraft lost and more than one hundred and forty damaged. The Group completed four thousand eight hundred and thirty-eight deck landings with only fourteen accidents. The last nine hundred and thirty-nine deck landings were completed accident free. Sixteen pilots had flown more than one hundred and thirty-nine sorties, with Lieutenant K. Whittaker RN achieving the record number of one hundred and forty-nine. Nine aircrew were killed and one wounded. Twenty-four aircrew had been rescued.

At the end of May, *Glory* again entered the dry-dock in Grand Harbour for a much needed clean up and refit during which full advantage was taken of the splendid facilities on offer in the island for rest and recreation. In July, resplendent in fresh paint once more HMS *Glory* wearing the flag of Vice-Admiral Edwards, Flag Officer, Second-in-Command Mediterranean Fleet, sailed eastwards from Malta and northwards through the Aegean Sea. In company with the Canadian aircraft carrier *Magnificent*, the cruiser *Cleopatra*, destroyers *Chivalrous*, and *Chevron*, they were going to Turkey via the Dardanelles to visit the chief seaport and former capital, the city of Istanbul, formerly known as Constantinople.

Unfortunately the visit was cut short after the third day by the redeployment of the Mediterranean Fleet in connection with the mounting tension in Egypt and the abdication of King Farouk.

Despite this, *Glory*'s visit to the 'Venice of the East' is worth putting on record. As the ships passed through the Dardanelles a running commentary was given over the ship's speaker system which took *Glory*'s crew back to the bloody fighting of Gallipoli, reminders of which still stand. On the port side stood the 70ft high British War Obelisk, and at the entrance a massive square stone Turkish fort. On either side of the ship could be seen smaller forts in various stages of decay. The different shades of green, varying from the dull shade of sparse grass on the hillsides to the lush green of the valleys, and sight of trees, were a welcome change from the sand coloured monotony of Malta.

On passing the entrance to the 'Golden Horn' harbour, *Glory* fired a National Salute of twenty-one guns, and further salutes were fired as the various Naval, Military, and Civil Authorities called on the Admiral. On the second day of the visit, Turkish and British Naval contingents marched to the Ataturk Memorial in Taksim Square, where the Turkish and British Admirals each laid a beautiful floral wreath at its foot. All the ships were open to visitors, but the numbers were necessarily restricted to the few hundreds who could come off in boats, as *Glory* was anchored about nine cables from the Custom House. However, they came off in sufficient numbers to keep the guides busy and the 'carriers' were a particular attraction.

The Turkish people were extremely interested in everything, and as is usual with ship visitors, there were always stragglers to be rounded up after the official closing time. Istanbul is a fascinating city, its history began in the year 658 BC when Greek colonists founded Byzantium. Since those days it has seen the Persians, the Crusaders, and finally the Turks, all of whom have made Istanbul the fine city it is today. The ship's company came away with memories of the Eastern beauty of the Museum of Avasophis (formerly the Moslem church of St Sophia), the rich carvings and sculptures of the Sarcophagus of Alexander the Great, the magnificence of the Mosque of Sultan Ahmet, and the splendour of the Dolmabahce Palace all of which were in view from *Glory*'s anchorage. There were so many splendid places to visit.

But a tourist by day, a sailor becomes after sunset the seeker of gaiety, bright lights, romance, dancing girls, and new exotic drinks. Places of entertainment were found and explored, some enjoyed drinking in the 'Park 'Otel', whilst others enjoyed a good meal in places like the 'Konyali', and others . . .? It was a pity *Glory* could not stay longer but it was not to be. The ship had embarked three Naval Air Squadrons prior to the Istanbul visit, 898 and 807 Squadrons were both equipped with Sea Furies, whilst 810 Squadron had Firefly aircraft. If the Egyptian situation had assumed a more threatening aspect they may well have been called upon.

Glory went to Tobruk, (where Vice-Admiral Edwards transferred his flag to HMS *Glasgow*), then proceeded to Cyprus to await developments. Situated in the north east corner of the Mediterranean, Cyprus, 140 miles long by 60 miles wide is the third largest island in that part of the world. Richard the Lion Heart passed this way and Shakespeare's Othello is related to the Phoenician occupation of the island. Cyprus is 'Lovers Island' where, according to legend, Aphrodite (the Goddess of love) was born of the foam where the sea breaks on the rocks off the coast of Paphos.

Glory's temporary abode was again the small town and seaport of Larnaca, the place where many of the younger members of the ship's company discovered the folly of drinking cheap brandy; one foolish lad jumped out of the 2300 liberty boat (in his number 6 suit) as it approached the port gangway, because he wished to go to the starboard one. Several people spent the night in those bare little cabins up for'd; two more, one a Leading Signalman, found that cycling and brandy didn't mix – they landed up in a very deep ditch!

Larnaca was a very quiet little place and ways had to be found to pass the time constructively. Exercise 'Jogtrot' was initiated to test communications, the objective being to rescue an imaginary Governor's daughter from an imaginary group of bandits. The daughter was saved and all communicators were given a recommendation by the platoon commander for some excellent communications. Football, hockey and sailing matches were organised by local teams ashore and the other ships in company. Cycles could be hired for one shilling and sixpence per hour and many ventured as far as Nicosia, an extremely hot, dusty and strenuous bike ride. Excellent swimming beaches were available, and drink was obtainable at fourpence a tot!

The one and only open air cabaret (The Goldfish Bowl) was very popular and everyone seemed to make it their final rendezvous for the evening. However, HMS *Glory* received 'further orders', the political horizon had cleared and a fond farewell was said to the birthplace of Aphrodite. Back in the familiar surroundings of the George Cross Island the three Squadrons that *Glory* had taken aboard were disembarked, and in turn the ship welcomed 801 Squadron, (Sea Furies) and 821 Squadron, (Fireflies) who were to return with *Glory* to the Far East and carry on the task in Korean waters. But, as a final fling in the Mediterranean came the visit to Barcelona, in September 1952.

Wearing the flag of Rear-Admiral Parham, Flag Officer, Flotillas Mediterranean, HMS *Glory*, with HM Ships *Chequers*, *Chieftain*, and *Chevron* in company, made history by being the first Royal Navy ships to visit Spain since 1934. The visit was a great success, the Spanish people are traditionally hospitable, and went out of their way to make the stay a tremendously happy one for the ships' companies. Barcelona, the capital of Catalan, is situated on the north east coast of Spain a country that has been described as, 'A land where life smiles under a perennial sun, a hospitable land, where the charms of the east combine with the comforts of the west, where the memories of a traditional past do not hamper the thriving, industrial present. It is a land full of contrasts and able to please the most varied of tastes'.

Short though *Glory*'s five day visit was, everyone could sense much of this and it was hoped that someday the ship would return to the lovely Iberian Peninsular. Many different peoples have passed through Spain. The Iberians, Phoenicians, Greeks, Romans, Visigoths and Muslims haved all left their mark and the Spanish people emerged from the melting pot, endowed with strong and very marked characteristics. This can be seen in the customs of the people, in the beauty of the countryside and cities, and in the immense artistic treasures to be found throughout the peninsular. No city could be more typically Spanish than Barcelona, with its huge column and statue of Christopher Columbus overlooking the harbour.

Barcelona, with its Plazas, bullfights, cabarets and its beautiful senoritas. *Glory* and the other ships were secured alongside within five minutes walk of the city centre; the Spanish people came visiting in their thousands and didn't want to leave, even being duty onboard was a pleasure. What could be more pleasant than showing a dark eyed Spanish beauty around an aircraft carrier? Another example of how pleasant a duty can be was on the third evening, when the show from the 'Baghdad' came onboard to entertain the duty watch and men under punishment, for who else was left on *Glory*? The flight deck abreast the 'island' was floodlit and the audience consisted of over two hundred men from all the ships in company. The floor show was made up of a Spanish orchestra, slim dancing youths in colourful Spanish dress, and dancing girls with tambourines, castenets, and their dark, flashing eyes. The applause was spontaneous and terrific, a mixture of cheers, handclaps and whistles, and the artistes loved it.

The enthusiasm put into this appreciation rivalled that of the dancers and certainly 'The Baghdad' had seen nothing like it. This indeed was a contrast to the flight deck scene of the Japanese Surrender in 1945, to the taxying of aircraft and the roar of engines. Daily visits were organised, the *Glory* boys went out in their hundreds and took in all that they could of this ancient, modern and most splendid city. Who would ever forget the panoramic view from the top of Tibidabo Mountain, with the whole city of Barcelona and miles of blue Mediterranean spread out below?

All were impressed by the magnificence of the Church of Sagrada Familia, which is the greatest monument of modern Barcelona. Many admired the beautiful paintings and art treasures in the ancient offices of Mayoralty, others enjoyed a ride up or down the Avenida del Generalissimo Franco, that wide highway dividing the city in two. Who did not enjoy a wander round the tortured streets either to see a bull fight, which was exciting, colourful and cruel; a demonstration of courage, elegance, grace and art, against the vital strength and power of the bull. Some obtained as souvenirs 'banderillas' actually used in the fights they saw, bloody, coloured paper covered darts. They cheered and booed the matadors with the crowd in the traditional Spanish manner; they even cheered the bulls.

Many of the matelots threw their caps into the ring at the end of a polished performance, which is an old Spanish custom. In the evenings everyone explored the night life of Barcelona. By going to bed, one would have missed much that was new and exciting, an opportunity that might never occur again in the life of an individual. Sitting at one of the tables on the wide pavements outside the numerous bars and cafes, one could drink wine or beer and watch a colourful world pass by. There was a mixture of caballeros, senoritas, and tourists from many different countries, it was easy to pick out the latter from the Spanish people. There were many cabarets, the 'Bolero', Emporium, Follies, Copacabana, and of course the 'Baghdad'.

There were many tales to recount and embellish. The Royal Navy lived up to its name of 'Ambassadors of the British Empire' except for one minor incident, when a sailor tried to insist that the statue of Christopher Columbus was really that of Sir Francis Drake, and an irate Spaniard convinced him otherwise to the tune of a broken jaw! Thus arrived the day of departure, when crowds and farewell scenes on the dockside were more reminiscent of a ship leaving its native port for a foreign commission than they were of an official visit to a foreign port. Anglo-Spanish relations must have been given a tremendous boost by the making and cementing of many friendships. *Glory* had certainly enjoyed the visit and so had her new found friends. The passage back to Malta was rough, and everyone was glad when *Glory* eventually slipped back into Grand Harbour. After a couple of days rest it was back to sea for more working up trials with the two squadrons and then back into harbour on Friday 3 October to take on fuel, food and ammunition in preparation for the voyage back to Japan. A last run ashore for the cherished steak, egg and chips washed down with coffee, all for three shillings and sixpence. Last letters home to Mum gave instructions that in future all letters shoud be addressed to 'HM Ships in Korean Waters' instead of 'Forces Airmail'. Then, at 1600 on Thursday 9 October 1952, *Glory* set off again for those Korean Waters. Life was still extremely tense in Egypt and when *Glory* secured in Port Said it was considered prudent to rig arc lights on booms in case of attack.

9

Completion of Korean Duties
(November 1952–July 1953)

There was still plenty of ribald banter though as the ship made her way along the Suez Canal on Monday. By Wednesday 15 October, *Glory* was well into the Red Sea with temperatures touching 107 degrees Fahrenheit! Water coolers were much in demand, and the ship's company gave a rousing cheer when it was announced on the BBC news bulletin broadcast over the ship's radio equipment, that snow had fallen in north Wales. In Aden on Friday and *Glory* received the eagerly awaited mail. Very few ventured ashore preferring to get some letter writing done before the long trek across the Indian Ocean. The next day *Glory* passed close to a small island, a leper colony, with its small houses and huts on the shore, and many of the inhabitants came out to see *Glory* steam past. On Sunday Socotra, a large island, was given close scrutiny by the ship's company as the ship went close in. It had some lovely beaches and high cliffs, but it looked totally uninhabited. However, there were RAF facilities for anti-submarine patrols and a population of about 10,000, somewhere!

The pilots of *Glory*'s new Squadrons had flown their aircraft from England to Malta in June 1952, but on the way an unexpected delay occurred in France when five out of ten tailwheels burst on the Firefly Vs of 821 Squadron, and replacements had to be sent from the UK. The Squadron was commanded by Lieutenant Commander J.R.N. Gardner RN, who had served as a naval test pilot at the Royal Aircraft Establishment Farnborough. The Firefly earned the nickname 'The Coal Burner', because of its slow speed and exhaust smoke.

The Firefly aircraft, however, proudly took their place amongst the formidable array of United Nations fighter/bombers operating over Korea. Perhaps the greatest asset of a Firefly was its remarkable deck hook which often appeared to grab arrester wires from all manner of awkward and dangerous looking situations. In fact

821 Squadron was to complete 1,487 landings without mishap, which showed just how skilled the pilots had become. 801 Squadron (Sea Fury), was commanded by Lieutenant Commander J.P.B. Stuart RN. The squadron had reformed in April 1952, at the Royal Naval Air Station Lee-on-Solent. One of their pilots was Commander B.C.G. Place VC DSC RN, who took part in the midget submarine attack on the German battleship *Tirpitz* in Altenfjord. Since returning from being a Prisoner of War he first qualified as a navigator and then entered naval aviation as a pilot in a front line squadron.

On Tuesday 28 October 1952 whilst *Glory* was on passage to Singapore, thirty aircraft from the squadrons swooped on the Kuala Langat forest reserve in south west Selangor, Malaya, to blitz a terrorist camp already under attack by artillery. Three raids were made at 0800, 1200, and 1500. The planes dropped 500lb bombs on the camp, and strafed the surrounding jungle with 20mm cannon fire. The *Glory* had interrupted her journey down the Straits of Malacca so that the attacks could be made as support for the 1st Battalion Suffolk Regiment and the Malay Police, who were making determined efforts to crush the remnants of the once notorious Kaj Ang gang.

Headquarters of 18 Infantry Brigade reported that, 'Preliminary reports say that the naval pilots were bang on their targets. However, it will be several days before we can confirm results, as the camp was deep in the jungle and it will take time for our ground patrols to get to it.' HMS *Glory* anchored between the mainland and Singapore island that night, whilst armed guards patrolled the decks again and floodlights were rigged on the waterline. A very strong 'buzz' swept the ship in Singapore, it seemed that *Glory* would sail up to Tower Bridge when she returned to the UK so that the coronation crowds could look her over! The ship's company were also going to get ten days' extra leave! After the short stop in Singapore, *Glory* proceeded to Hong Kong and took over from HMS *Ocean* on Tuesday 4 November.

As usual the first stop on going ashore was the China Fleet Club, where the menu had a hundred and ninety items on it, practically every kind of food a matelot could wish for, served up very quickly and cleanly. Half a lobster with chips and all the trimmings for three shillings, or lemon sole, chips, and coffee for one shilling and ninepence. It was also a joy to sit down at a nice clean table with excellent service. The troopship *Empire Fowey* was given a cheer when she left for the UK the next day, packed with soldiers from Korea who would be home for Christmas. An addition in the Air Petty Officers' mess was a red parrot, who walked up and down the mess table, knocking over anything in its way. It would perch on a rail, balancing on one claw, while eating food out of the other.

The weather in Hong Kong had been warm when the ship's company had been issued with their winter clothing which included fur hats and gloves, but less than a week later they were very glad of it. *Glory* arrived back in Sasebo on Sunday 9 November, and left the following day for the operational area where the familiar routines began. Long before dawn the work of preparing the aircraft for the day's sorties would start. In the murky darkness with the ship rolling and pitching in the rough seas, the fitters and riggers, electricians and armourers went about their tasks. Below in the briefing room the aircrews were being informed of their targets, routes, identification signals, weather, and enemy activity. Then out of the cosy fug, up onto the freezing inhospitable expanse of the flight deck, stumbling over the wire barriers.

Then suddenly, the crack of koffman starter cartridges as they fired the aircraft engines into life. Pilots tested controls, the roar as engines were boosted, and then in the dim light of early morning the shadowy figures of the aircraft handling party, crouching low as they scurried towards the safety of a sponson on the edge of the flight deck clutching the heavy wheel chocks, desperately trying to stay on their feet as a tempest howled along the deck threatening to blow them into the whirling, almost invisible propellers of the planes as they taxied forward onto the catapult for launching. A final crescendo as the catapult hurled a laden aeroplane along its sixty-five yard track and over the bows into a grey maritime dawn.

The patrol was hampered by the miserable weather but even so the usual targets of rail-bridges, tunnels, ox carts and lorries were attacked. On Thursday to the great delight of everyone aboard, mail arrived by air, the first for ten days. Then tragedy struck on two consecutive days. On Tuesday 18 November, Lieutenant Richard Nevill-Jones RN was killed whilst carrying out an attack, and the following day Petty Officer (Air) Victor Coleman was blown off the flight deck by the slip-stream of an aircraft, he sank immediately and no trace of him could be found despite a long search by the helicopter. On the same day *Glory* closed up to defence stations on receiving an 'imminent' air raid warning, which had everyone searching lockers for long forgotten life belts.

The alarm quickly passed and the ship made ready to leave the operational area. Two 801 Squadron pilots, Lieutenant P. Wheatley RN and Lieutenant R.J. McAndles RN, had the unique experience of being transferred by jackstay to the destroyer HMS *Comus*, which then took them to the American aircraft carrier *Badoeng Strait*, CVE 116, (also known as the *Bing Ding*) where they observed flying operations during that ship's patrol on the west coast. *Glory* arrived in Sasebo on Thursday afternoon. To make a change from the usual run ashore in Sasebo, Petty Officer W.G (Glyn) Thomas, organised a coach trip to a small town about thirty miles away. The coach cost £6 to hire and thirty Petty Officers were transported in comparative comfort to Ureshimo, which was attractively situated in the mountains and had natural hot springs.

It was an extremely tranquil town, not the place that 'jack ashore' is usually associated with, and the local inhabitants were truly amazed at this sudden appearance of the Royal Navy, and it did not take long for word to spread! Small children were soon clustering round to stare and point. The local hotel was overjoyed at the unexpected boost to its takings when everyone decided to have a meal after the sight-seeing. The food was good and cheap and made all the more interesting by being served on tiny tables by pretty Japanese girls, whilst their guests squatted on the carpet. The hotel has probably never again witnessed the sight of thirty pairs of British shoes lying at the entrance to their establishment. It was a most enjoyable outing which was rounded off nicely by a visit to the Navy Club in Sasebo to see the floor show.

It was whilst in harbour that pictures were received from the 'Daily Graphic' whose photographer had been aboard during the passage from Singapore to Japan. *Glory* left Sasebo on Friday 28 November to commence the second patrol, but it was a very uncomfortable start. The ship encountered a gale during the first day and began rolling and pitching in an alarming manner. Only those essential for the running for the ship turned to, while the rest crept thankfully away to bunks and hammocks. On Sunday, Captain Maunsell, who had become very ill, was put in a stretcher and transferred to HMS *Consort* by jackstay and taken to Sasebo. Commander Bromley-Martin RN assumed temporary command of *Glory*.

Snowstorms and a bitter north easterly wind combined to cause problems with aircraft serviceability, and personnel working on the flight deck were encased in a wide range of clothing in an effort to remain warm; waterproof sheepskin lined coats fastened with metal clasps, big fur leather caps, golfing jackets, thigh boots, and a variety of old school scarves. Despite the weather, over three hundred sorties were made for the loss of one aircraft from which the pilot was rescued unharmed. The ship began the journey back to Japan on Monday 8 December, this time to Kure, and after the recent atrocious conditions the trip through the Inland Sea was a delight.

In this peculiar war in which *Glory*'s aircraft were employed almost entirely in supporting the Army ashore, the selection of targets was made by Army personnel embarked in *Glory*, known as 63 Carrier Borne Ground Liaison Section. This small group of 'Pongos' (three officers and three other ranks) was responsible for training *Glory*'s pilots in all aspects of army support. This entailed map reading, tank recognition, bombardment spotting and allied subjects. Intelligence reports, pilots' debriefs and photographs all had to be sifted. Records had to be maintained, and all material had to be indexed and classified. To give an idea of the work involved, over four thousand photographs of enemy territory were developed 'on an average' for every patrol of nine days. Much useful information was produced from these photographs, enabling successful sorties to be carried out.

The Carrier Borne Ground Liaison Section worked in close co-operation with *Glory*'s own Photographic Section and the Photographic Interpreter. The principal targets in Korea were all forms of transport whether road, rail, or waterborne; coastal gun positions; troop and supply concentrations, and electrical installations. The enemy was extremely cunning in his use of camouflage, but with experience *Glory*'s pilots were able to detect many targets while on reconnaissance. During patrols the officers of the section briefed and de-briefed all strikes against ground targets. 63 CBGLS certainly played its part well in *Glory*'s operations and contributed much to the good relations which must exist between services in peace and war.

Captain E.D.G. Lewin, DSO DSC* RN arrived to take command of HMS *Glory* on Sunday 14 December 1952, flying out from the United Kingdom. The ship left Kure the next day for her third patrol. On Tuesday it was decided to let newly joined pilots carry out deck landing practice. All went well until the afternoon, when the helicopter was being manoeuvred into position for'd of the island. It lifted clear of the deck, but unknown to the pilot the nose wheel steering arm was still attached. The aircraft appeared to gain height but the steering arm hit the jib of the mobile crane parked in front of the island causing the helicopter to spin, the tail rotor then hit the bridge and disintegrated, and the helicopter plunged into the sea. The pilot, Lieut Alan Daniels DSM RN and Chief Petty Officer E.R. (Ted) Ripley, were both killed.

Lieutenant Daniels had joined the Royal Navy at the age of sixteen. In 1942, while serving in the Mediterranean in the aircraft carrier *Argus*, then on convoy duties, he was awarded the Distinguished Service Medal. He lived in Bristol and left a wife and two children.

The weather was better for this patrol and flying was restricted on one day only due to snowstorms, but two more aircrew were killed. On Saturday 20 December, Lieut Peter Fogden RN crashed into the sea, and on Christmas day, Lieut Robert Barrett RN failed to recover from his dive after attacking railroads. It was a very sad ending not just to the patrol, but to a year's hard work. It was a subdued ship's company that returned to Sasebo on Sunday 28 December.

The 'Daily Mail' had provided the ship's company with Christmas gifts of cigarettes, 'nutty', and beer, which helped to bring some festive cheer, but what gave the ship's company most satisfaction was the party they gave to the children of Sasebo's orphanage on New Year's day, when the ship echoed to the delighted squeals of the very small children all afternoon. The miserable weather and the operational requirements had a drastic effect on games like volley ball and deck hockey, but many other pastimes were pursued by officers and ratings. Hobbies ranging from embroidery and woodcarving, to rug making and painting were taken up. Reading became the most popular occupation, and the three librarians were kept busy by a ship's company who read a huge number of books. Magazines and newspapers were also obtained from home.

There were some aboard who spent their leisure time studying by correspondence course, and subjects included Law, Languages and Electrical Engineering. There was also a keen French Group, and one rating managed to make some progress teaching himself Russian! Formal education was available at all times and many of the ship's company passed the Higher Education Test, Education Test 2 or Education Test 1. Bandmaster Pearce of the Royal Marines provided a series of weekly concerts in the chapel and this recorded classical music drew a regular group of keen listeners. Other musical programmes were provided by the ship's radio engineers who also broadcast the news, Padre's corner, the inter-part quiz, and individual mess programmes.

Glory began her fourth patrol on Sunday January 4 1953, and operations commenced next morning. It was a day of drama and tragedy. In the morning a Sea Fury made a forced landing after engine failure, and a short time later Lieut Mather RN was seen to bail out of his aircraft. Two aircraft were dispatched to act as escort to the helicopter which went to locate the missing pilot, but low cloud hampered the search and tragically Sub-Lieut Brian Rayner RN was killed when his aircraft crashed. Later that day Sub-Lieut James Simonds RN was also killed when his aircraft hit the ground after carrying out an attack. Notification was later received that Lieut Mather had landed safely but was taken prisoner. The next day two American admirals were on the flight deck when Lieut Wilfred Heaton RN was brought back to *Glory* after being rescued.

Lieut Heaton had been forced to ditch his aircraft having been hit by small arms fire. A US Airforce helicopter rescued him in record time and on arrival onboard *Glory*, the chopper pilot was thanked by Captain Lewin and presented with a bottle of whisky. The ship left the operational area and began the passage to Kure.

During *Glory*'s many patrols the 'jackstay' transfer had become such a prominent feature of everyday shipboard life that a special bugle call was introduced to summon the 'jackstay party'. Several hundred transfers were made, carrying everything which goes to make up naval life, from admirals to ordinary seamen, ammunition to potatoes, bread and mail, the latter being the most welcome and essential item. It was estimated that more letters and parcels arrived in the ship this way than 'postie' carried over the gangway.

For VIPs, a special chair was constructed and they literally became 'chairborne' while in transit between two ships. Many too were the nationalities of the ships with whom *Glory* carried out transfers; British, American, Canadian, Australian, New Zealand, Indian, Pakistan, and Dutch – a United Nations effort in reality! At first the samson post aft was used, but this proved far too long an operation, since it had to be rigged, then stowed away after each transfer, besides which the work on aircraft was often hindered. Then a method using the crane was adopted which proved much quicker as the gear could be left partially rigged. The time taken to rig completely was three minutes, the old way took thirty minutes.

A trial run of ammunitioning at sea with the Royal Fleet Auxiliary *Fort Sandusky* was carried out with a wind of 35 knots over the flight deck with intermittent snow showers, and a good rate of transfer was achieved. Shortly after this HMS *Newcastle* brought *Glory* some badly needed 20mm ammunition from Hong Kong and this transfer was a terrific effort. In atrocious weather, even worse than the previous occasion, but with no fewer than three jackstays rigged, nine hundred boxes of ammunition were transferred in just over an hour. The job was actually completed with little or no interference to the flying programme. The dreary routine began again on Monday 19 January, when *Glory* left Kure for the fifth patrol. The weather again was dreadfully cold and some aircraft were even sent ice spotting in case the enemy decided to go island hopping across the ice!

But at least there were no casualties on this trip and it was back to Sasebo on 29 January, where the exceedingly tedious business of victualling and storing ship would have to be endured. Small gangs of men bowed down under cases of condensed milk, sacks of spuds, boxes of soap, jam, fruit and herrings, would wend their way down flights of iron ladders, through narrow hatchways and passageways, until thankfully they would reach the remote storerooms deep in the bowels of the ship. Petty Officers and Leading Hands would desperately try to keep a tally of each case, box or sack, but there was always the lucky mess who would benefit from a 'broken' box of tinned fruit or other goodies which became lost during the tortuous trip. A modern aircraft carrier equipped with weapons of terrifying accuracy had to be reprovisioned using methods employed by ancient Egyptians! As the lower deck lawyer said, 'It keeps the troops out of mischief mate.'

Living in a tiny area teeming with bodies requires personal cleanliness of a very high order, there are no two minds about it. If it were necessary to bring pressure to bear on any member of the mess there would be no need to seek assistance from the hierarchy – the messdeck would provide it! A concern for personal appearance was a quality that jolly jack could rightly boast of as he fell in to be inspected by the Officer of the Watch before going ashore. A typical sailor would be a picture of sartorial elegance, with the bow of his cap-tally teased into a flat symmetrical rosette; a spotless blue jean collar, its white borders gleaming; the black silk ironed to impeccable smoothness; the ribbons attaching it to the jumper were as long as possible with swallow tail ends; the bell bottoms as wide as could be achieved and pressed concertina-wise to form seven horizontal creases on each leg. It was a relief and pleasure for *Glory*'s ship's company to get ashore and enjoy a good meal and drink of beer from a steady table after the days spent carrying out monotonous tasks in a damp, cold, dangerous and miserable environment.

The *Glory* went back to sea on 5 February, to resume her duties on the west coast. There was rather an unusual incident on this patrol when officers of the Royal Fleet Auxiliary *Wave Knight* became so unwell they were unable to bring the ship to refuel *Glory*, and officers from the cruiser HMS *Birmingham* were seconded to the tanker, which arrived only one day late.

On Monday 9 February, Sub-Lieut Millett RN and Captain Ralph Berry (Royal Artillery) one of *Glory*'s Carrier Borne Ground Liaison officers, had some amazing good fortune. Their Firefly aircraft swung sharply to starboard after landing and plunged into the sea. 'There was I, at minus fifty feet, trying to bale out of the plane' said Captain Berry. It was an experience that he could jest about, but it was one that nearly led to his death fifty feet below the surface of the icy sea off Korea. Captain Berry said, 'The plane sank very quickly and Sub-Lieut Millett reacted quicker than I did, getting out just as it sank. He saw that I had not got clear and dived down to help me, but the plane was sinking too fast.

'As we hit the water I had partly opened my canopy and as the plane was sinking on its back the cockpit immediately filled with water. I took a deep breath and struggled to jettison the emergency hatch. My ears were popping with depth and my lungs were bursting before I got it free and struck out for the surface. I thought I would never make it, I had torn my immersion suit as I struggled free and it became waterlogged and tended to drag me down.' A whaler from the destroyer *Comus* had been launched and was speeding to the spot. There was still no sign of Captain Berry. One minute and twenty three seconds had passed since the plane had disappeared beneath the waves. 'My lungs could stand it no longer', continued Captain Berry. 'I took an involuntary breath and found myself breathing air, I had made the surface.' But his troubles were by no means over, the rent in his immersion suit designed to keep him afloat and keep out the numbing cold, made it valueless. For the next three minutes he fought against the cold and the drag of his saturated clothing. 'Three minutes after I had surfaced I was dragged aboard the whaler but it seemed like a life time', he added. Safely back aboard *Glory*, Captain Berry was treated for minor cuts, immersion, and exposure, but he was back on duty next day. It was not the first occasion that Ralph Berry had crashed into the 'drink'. During the second world war, serving in a similar capacity aboard another British carrier he crashed into the sea off Burma. 'The sea there was a lot warmer than Korea,' he said.

The horror of this wretched war was again brought home to all on *Glory* on Wednesday when Lieut Cedric MacPherson RN was killed whilst carrying out a low level attack, and on Saturday, Sub-Lieut Richard Bradley died after his aircraft was forced to ditch. The ship arrived back in Kure on Tuesday 17 February to learn that no tugs were available to assist in going alongside, so Operation Pinwheel was put into effect. This entailed using the slipstream of a number of aircraft to manoeuvre the ship alongside. It was a very effective evolution but caused a certain amount of anxiety to the aircraft engineers because it required their precious aero engines to be boosted to high revolutions with the danger of overheating.

So ended the sixth patrol. In Kure the opportunity was taken to give those personnel who were taking promotion examinations the chance to exercise their 'power of command' by carrying out squad drill on dry land, instead of trying to navigate teams of matelots around a busy and congested flight deck. The run ashore in Kure now offered lots of modern clean shops with flashing neon lights and a large choice of goods for sale. The 'Kure Club' was serving the very popular and appetising menu of rump steak, two eggs, and of course chips for just three shillings, and there was also the novelty of a glass of 'fresh' milk.

The usual atrocious weather was encountered on the next two patrols with high winds, rain, snow and fog. The early morning fogs in the Yellow Sea did benefit those engaged in flying operations in that the first sortie usually could not take place until about 0900.

But for the destroyer escorts experiencing nasty seas with boisterous steep sided waves, living conditions must have been extremely hazardous as their vessels plunged wildly into the deep troughs. Generally conditions above and below decks were very miserable during these first few months of 1953. It seemed that every mess deck and cabin had its share of damp clothing hanging around, adding to the discomfort and irritation of overcrowded living spaces. The smell of grease, sweat, steam and tobacco smoke; the shouting, singing, talking mass of jumbled bodies, the ship's radio valiantly trying to compete with the cacophony, at times drove even the hardiest to seek peace and quiet. So down in the labyrinth of alleys and passages solitary figures would settle down to read or write for a couple of hours.

Down on the canteen flat was the NAAFI, with its village shop aroma of soap, chocolate, and that most important of commodities, Johnsons Baby Powder (Foo Foo). At opening times it was an oasis where tins of pears or peaches and Nestle's cream could be bought for a couple of bob, to improve the flavour of the figgy duff.

Nearly every day whilst on patrol *Glory* steamed three hundred miles, and it was disclosed that up to 31 January 1953 the ship had steamed over one hundred and thirty thousand miles since leaving the United Kingdom. There had been seven thousand operational sorties over Korea. This number of flights could not have been safely carried out had it not been for the Safety Equipment Section, who packed and serviced one thousand four hundred parachutes, two of which were used to make descents. Innumerable life jackets and dinghies were packed, forty of these being used in earnest.

After the seventh patrol *Glory* had returned to Sasebo on Sunday 8 March, leaving again on the following Sunday. When the ship returned to Kure once more on completion of the eighth patrol, on Thursday 26 March, HMS *Unicorn* was waiting with replacement aircraft. The *Unicorn* had been in the Far East since the Korean war started and had played a magnificent role in supplying and repairing aircraft for the light fleet carriers. She had also taken part in shore bombardment in support of land forces, her 4in guns pounding enemy positions. Her other duty was to ferry troops.

During this stay in Kure a memorial service for the late Queen Mary was held on the jetty between *Unicorn* and *Glory*, on Wednesday 1 April. With just the slightest hint of better weather on the way the ship's company were now thinking increasingly of going home. Everyone was weary of the war, it had been a long bleak winter which generated unwelcome memories for the old Russian Convoy hands. Bitterly cold, visibility poor enough 'to make the seagulls walk', and ice – ice in sheets and ice in packs solid enough to hazard a destroyer's thin skin. And whereas even the most junior hand knew why he was fighting World War Two, Korea was an alien place containing someone else's war, a place with little more substance than an outline seen faintly through a passing snow squall. What enthusiasm there was, was given by the knowledge that the mixed United Nations battalions had it far worse, shivering in their foxholes with their backs to the sea and at times outnumbered and totally dependent upon carrier borne air support.

On Friday 3 April, *Glory* left Kure for the ninth patrol. The weather was certainly improving and as if to celebrate the arrival of spring, on Easter Sunday 801 Squadron and 821 Squadron flew a record one hundred and twenty three sorties, it was the culmination of brilliant team work for which *Glory* had become renowned. It was this proficiency and skill which had enabled *Glory* during her time in Korean waters regularly to launch aircraft twice as quickly as the United States' carriers, despite having only one catapult.

Amidst all the frantic activity, full services were held in the ship's chapel, and the helicopter was busy transferring other chaplains to the escorting destroyers so that divine services could be held. It was a truly memorable Easter Day. The ship arrived back in Sasebo on Sunday 12 April 1953, with everyone feeling pleased with the knowledge that they had contributed or been witness to another milestone in naval aviation. The following week *Glory* set off for her penultimate patrol. The usual selection of targets were successfully attacked, and all went well until the evening of Saturday 25 April, when it was learnt that Lieut John McGregor RN and Sub-Lieut Walter Keates RN had both been killed after their aircraft had crashed following attacks on enemy positions. It was such a cruel blow, to have happened so near the end of the tour.

Admiral Sir Charles Lambe, who during the second world war had commanded the aircraft carrier *Illustrious*, paid a visit to *Glory* and flew on a sortie over the enemy coast to see for himself how efficiently the aircrews operated. The ship returned this time to Kure, and secured alongside on Wednesday 29 April. A very popular purchase for many of the ship's company was a lovely Japanese tea set, which was exquisitely decorated in gold paint with scenes of Mount Fujiyama, and a real bargain at around £3!

HMS *Glory* began her final patrol on 5 May, and as she steamed through the Inland Sea for the last time many of her company made their way to the flight deck to view the magnificent scenery.

Before operations began aircrews were advised that no unnecessary risks should be taken whilst carrying out attacks! This rather belated advice had come about in response to some recent losses suffered by the United States' carriers, but in *Glory*'s case seemed superfluous. As it was, fog and haze played its part in delaying operations. However, finally on Thursday 14 May, the last operational sorties from *Glory* were carried out. The weather remained fine all day, and at 1415 the final aircraft was recovered. It was all over. HMS *Glory* and her ship's company had acquitted themselves admirably during the three tours in these alien waters, and would now be able to start the passage home to where the final preparations were being made for the Queen's Coronation in June.

The 'buzz' that *Glory* would steam proudly up the Thames to Tower Bridge had long since died, and one newspaper printed a story claiming that the ship would be held over in Singapore because it would be too dirty to be among the spotless ships at the Spithead Review. None of which really worried a ship's company who were only anxious to get home. On Sunday 17 May 1953, *Glory* arrived in Sasebo and began handing over to HMS *Ocean* again. At last on Tuesday, with bands playing and with the cheers of thousands ringing out across the harbour, HMS *Glory* left Japan and began the voyage home. It seemed an interminable journey. At each port there were vehicles, crates, containers and packages to be stowed away and lashed down, and there were also service personnel to be embarked for passage to the United Kingdom.

Coronation Day was celebrated in Singapore in the traditonal manner with 'splice the mainbrace' and the ship dressed overall with flags and bunting. There was the familiar sweltering journey through the Red Sea. But on Friday 3 July there was a last opportunity to enjoy a run ashore in a 'foreign' port – Gibraltar! Early on Sunday morning *Glory* began the last leg of her long voyage. At 1000, lower deck was cleared for divisions and presentation of Long Service and Good Conduct medals. Captain Lewin took the opportunity to thank the ship's company for all their splendid efforts whilst carrying out an arduous commission and wished everyone a very happy leave. Divine service was held on the Quarter Deck.

On Monday 6 July general payment was made, and the excitement on messdecks and in the wardroom was almost intoxicating and that was before tot time! Suitcases and kitbags were being packed with all the treasures of the Orient, and the end was in sight. A crowd estimated at nearly two thousand was on the wharf at Portsmouth Dockyard when *Glory* flying a very long paying off pennant berthed at North Corner Jetty just after 1000 on Wednesday 9 July 1953. Her arrival was greeted with thunderous cheering and waving of handkerchiefs and emblems from the dockside. Many of the relatives and friends had been waiting since early morning, and to cater for them a special canteen had been set up on the jetty. A Bluejackets Band also played whilst the crowd waited, and as *Glory* drew near the strains of 'Home Sweet Home' drifted across the water.

Among the crowd were many brides-to-be who had travelled from distant parts of the country to greet their fiancés. *Glory* had arrived at Spithead earlier in the morning and was boarded out there by Vice-Admiral J.A.S. Eccles (Flag Officer Air Home), who gave an official 'welcome home' to the ship's company. The crowds on the jetty were the largest seen since the aircraft carrier trooping operation which followed the end of the Japanese war in 1945–6. An hour after *Glory* had berthed people were still being admitted to the Dockyard and making their way to North Corner Jetty, which for many hours was the scene of happy reunions.

First to welcome *Glory* home though was Mrs Mary Cardwell, of Coverack, on the Cornish coast. She stood at a bedroom window of her cottage home and with her four children waved towels and sheets as the carrier rounded the Lizard and passed close by to anchor three miles out from Falmouth. On the bridge of *Glory* waving back to them was her husband, Lieutenant Commander Albert Cardwell.

Sixteen customs men had boarded the ship off Falmouth to complete the Customs clearance so that the ship's company could go on leave as soon as *Glory* berthed in Portsmouth. Between £4,000 and £5,000 duty was paid on gifts that had been brought home for relatives and friends. As soon as the for'd brow was in position matelots and Royal Marines literally tumbled over themselves to get down to the jetty. One rating carrying three big suitcases relied on his burden to support him, and he arrived at the foot of the brow neatly sandwiched between them.

Just over an hour after *Glory* berthed the ship's padre, the Reverend Richard Knight, was at work again. This time to christen the nineteen month old daughter of Petty Officer George Hill, of Hanwell in Middlesex, who was seeing her for the first time. Maria, his twenty-one year old wife, clutched her husband's arm and whispered, 'The christening was the most thrilling moment of my life.' The ceremony was performed in an upturned ship's bell in *Glory*'s chapel, and the baby was named Lynne Maria GLORIA. The Padre said, 'Romance kept me busy announcing the banns for at least thirty-five Church of England sailors. Goodness knows how many other denominations there were, I couldn't keep count. I had to call all the banns three times, and was almost repeating them in my sleep!'

One of the *Glory* men engaged to be married was twenty-two year old Lieutenant Keith Sheppard, of Chester. He said, 'I feel it's about time I dropped anchor somewhere.' His fiancée was twenty-year old Ann King, from Gloucestershire. Leslie Clifford, a twenty-three year old Electrician's Mate from Bourton-on-the-Water, Gloucestershire, began a penfriendship with twenty-two year old Iris Hubbard while *Glory* was in Korea. Iris – who came from Priter Road, Bermondsey – had received more than a hundred letters from Leslie, but they had never met. On the jetty she was wearing identification clothes, a black coat, white gloves, and white close fitting hat. Leslie said, 'She has sent me a photograph. She's a lovely girl. Her letters are so homely too.' Marriage? 'Well you never can tell.'

Also waiting was eighteen-year old Olga Clarke, the friend of Royal Marine Harry David Phillips, aged twenty, of West Harnham, Wiltshire. She said, 'I got a job in a factory making ammunition at £3 5s 0d a week. I did it to help shorten the war and speed up the romance.' Wherever twenty-four year old Dennis Parkes went, his twin brother Tom was at his side. For Dennis, home to marry Jean Waterman, aged nineteen, it meant wedding bells within a few days. Tom, like his brother a stoker mechanic, would marry Vesta Hook aged twenty. 'It was love at first sight for both of us', Dennis said. Also to come ashore from *Glory* that day was 'Bertah', she was a honey brown Malayan bear, the gift of General Templer to London Zoo. Her 'dietician' on the way home was Lieutenant Peter Hanscomb, of Brighton. Bertah's daily diet was half a pound of rice sweetened with condensed milk, two raw eggs, an apple, a pound of mashed potatoes and six ounces of treacle.

Since leaving the UK in January 1951, *Glory* had spent 530 days at sea and steamed 157,000 miles. She had completed 15 months of war service and there had been a total 13,700 flights from her deck, 9,500 of which were operational sorties over Korea. Her ship's company had endured two Korean winters when ice and snow had to be repeatedly cleared from the flight deck and her aircraft, but despite these conditions operations had been sustained. The ship's company could look back with pride on these magnificent achievements.

10

Bowing Out
(October 1953 – August 1961)

The tumult and the shouting dies;
The Captains and the Kings depart.

On Saturday 25 July 1953, seventeen days after that glorious homecoming *Glory* left Portsmouth for Rosyth to undergo a much needed overhaul and refit, and secured port side to at the South Arm promptly at 0800 on Monday. The ship was next assigned to duty with the 1st Aircraft Carrier Squadron, and proceeded to Portsmouth on Monday 26 October 1953 to take on stores and embark the personnel from 826 and 801 Naval Air Squadrons, who had been based at Lee-on-Solent. The aircraft, Fireflies of 826, and Sea Furies of 801, were landed on soon after the ship had left Portsmouth on Monday 2 November. *Glory* then called in at Plymouth on Thursday, and proceeded to Gibraltar the following day. Only a few hours were spent in Gibraltar Bay, then it was back to an intensive flying programme all the way to Malta, where the aircraft and Squadron personnel were disembarked to the air station at Hal-Far.

For the remainder of November and all of December *Glory* went to sea during the week and returned to Grand Harbour for the week-end. Everyone enjoyed the Christmas festivities including Naval Airman Bernie Cohen, who had previously served in *Glory* from September 1949 until October 1951, this was his second Christmas in Malta aboard the ship, and during his naval career Bernie spent over three years aboard. It was back to sea for more flying before the old year was out.

In January 1954, *Glory* visited Naples for five days and in February went to Villefranche, on the Cote d'Azur where many visited the lovely towns of St Tropez, St Raphael, Nice and Cannes. Between the times many hours of flying were accomplished, taking part in exercises with ships of other navies. But

100

Glory's days as an operational aircraft carrier came to an end when the two squadrons were flown off to Lee-on-Solent on Sunday 28 February 1954 and *Glory* secured alongside in Portsmouth the following day. The ship's next role was as a ferry carrier, transporting all the necessary military paraphernalia needed to safeguard the few remaining outposts of the British Empire. In August, army vehicles, stores and a large number of 'cocooned' naval aircraft were loaded aboard the ship in Portsmouth Harbour.

There was another charming little ceremony in *Glory*'s chapel on Sunday 29 August, when the daughter of Lieutenant (E) Fitzgerald RN was christened. Before sailing for Glasgow on 13 September, about fifty Chinese cooks and stewards were embarked for passage to Singapore, and other Royal Naval personnel also came aboard for passage to naval ships and establishments in the Mediterranean and Far East. When the ship entered the King George V Dock in Glasgow on Wednesday, the weather had begun to deteriorate, and by Thursday, the day *Glory* was due to leave, the wind had increased to severe gales preventing the ship's departure. By Friday at 1600, conditions had eased enough to allow *Glory* to leave the Clyde and head for the Far East.

Personnel on draft to HMS *Rooke*, the naval base in Gibraltar, were landed on Tuesday 21 September, after which *Glory* headed into the Mediterranean. This time there was none of the hustle and bustle associated with flying stations, no aircraft engines roaring at full power, no screech from the catapult as it hurled the aeroplanes over the ship's bows, none of the frenetic activity as the flight deck parties ranged the aircraft. Those onboard were enjoying almost cruise like conditions on the thousand mile passage to Malta. The old familiar bars in Valletta and Floriana, The Bing Crosby, The British Empire, The Rose of England, the Egyptian Queen, were still doing a roaring trade, with Minco, Cookie, Sparrow, Bobbie, and Frankie, in charge of the entertainment. The experienced hands were delighted to search out the old haunts whilst the uninitiated sprogs were amazed at the glorious and famous Gut.

From Malta, *Glory* continued to Port Said, where since June 1953, when Egypt had become a republic, the attitude of the local citizens to jolly jack had become distinctly frosty, and it was therefore with some relief that *Glory* entered the Suez Canal at

2340 on Tuesday 28 September. Late the following morning the ship anchored in the Great Bitter Lake to await the arrival of other ships who would make up the convoy to go through the next portion of the Canal. It was a fine sunny afternoon and hands were piped to bathe from the port side at 1630. Next morning, *Glory* ran aground on mud at 1230.

Two tugs, *Beetle* and *Fly*, were called out to take up the tow with eight inch manila line, but to no avail, the ship had stuck fast. At 1420 the Motor Lighter *248*, which was attached to the Royal Engineers, also attempted to tow *Glory* off the mud but was unsuccessful and at 1445 the attempt was abandoned. The tugs *Beetle* and *Fly* were now joined by the tug *Vigilant* and further attempts were made to release the ship. For two more hours all three tugs did their utmost to release *Glory*. At 1700 the three tugs were joined by another, the *Atlas*, all four concentrated their efforts and eventually at 2055, after being held fast for eight and a half hours *Glory* slid free much to everyone's relief. An examination of the ship's hull by divers could find no damage and the ship was able to proceed into the Red Sea.

There was the usual short stop in Aden to take on water and fuel, and then across the Indian Ocean to the lovely harbour of Trincomalee, Ceylon. The island had been under British rule since 1815, and the Royal Navy had increasingly built up its facilities. Ceylon had become a self governing state in 1948, but the naval base situated in one of the world's loveliest natural harbours was still kept extremely busy. *Glory* arrived there at 0730 on Tuesday 12 October, when the usual diplomatic courtesies were observed as the local dignitaries came aboard to pay their respects, and *Glory*'s commanding officer, Captain H.W. Sims-Williams RN, returned the calls accompanied by the ship's senior officers later in the day.

The ship left the 'Resplendent Isle' on Friday, for the five day trip to Singapore dockyard, the last port of call on the outward voyage. During the ten day stay the ship's company were able to enjoy the facilities on offer at the Royal Naval Barracks (HMS *Terror*), and many went ashore to visit Singapore city, to sample the delights of Lavender Street and other haunts frequented by sailors from all over the world. On Tuesday 26 October Admiral

Lady Cynthia Mary Brooke launching HMS *Glory*, Saturday 27 November 1943
Musgrave Shipyard, Harland & Wolff, Belfast

HMS *Glory* on the slipway, Saturday 27 November 1943

Barracuda aircraft launched from HMS *Glory* off Scottish coast
April 1945

Barracuda aircraft undercarriage collapse
HMS *Glory*, April 1945

Barracuda aircraft crash landing, HMS *Glory*, off Scottish Coast
April 1945

HMS *Glory*
Fatal crash Corsair (Able Seaman Thompson) Friday 1 June 1945

HMS *Glory*

Admiral Jin Icha Kusaka, General Imamura, General Sturdee and Brigadier Sheehan with Captain W. S. Buzzard RN

Captain's cabin, Thursday 6 September 1945

HMS *Glory*

108

HMS *Glory*
General Imamura signs Surrender Document, flight deck, Thursday 6 September 1945

HMS *Glory*
Released prisoners of war, bedded down in the hangar

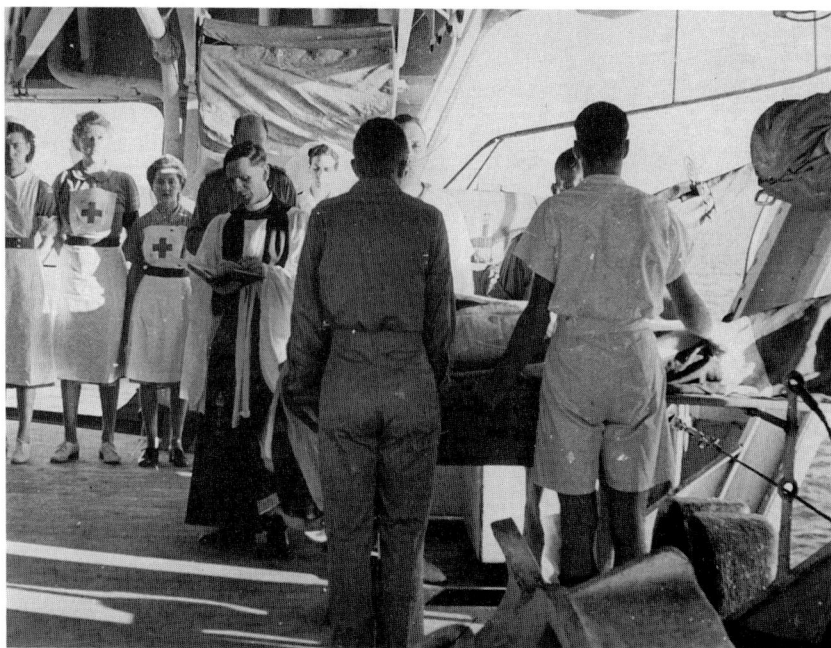

HMS *Glory*
Funeral of Sapper William Owens, from the quarter deck
off Eniwetok Island in the Pacific, Monday 15 October 1945

HMS *Glory*
Corsair and Barracuda aircraft, off Ceylon, July 1945

HMS *Glory*
Trincomalee Harbour, Ceylon, August 1945

HMS *Glory* coming alongside Station Pier, Port Melbourne
HMS *Indefatigable* in right background, Wednesday 23 January 1946

HMS *Glory*
Ratings from *Glory* march past Saluting Base
Swanston Street, Melbourne, Friday 25 January 1946

HMS *Glory*
Stern view of *Glory*, Captain Cook Dry Dock, Sydney, Australia, 1946

HMS Glory

HMS *Glory*
Corsair landing, 'Bats', June 1945

HMS *Glory*
A Manchurian crane which rested on the flight deck, March 1952

Mr David Wharton presents memorial seat to people of Grenfell
Australia

HMS *Glory*
Corfu, Greece, 1950, Italian training ship *Amerigo Vespucci*

HMS *Glory*

HMS *Glory*
Lieut-Cmdr F.A. Swanton making the 2,000th deck landing since leaving the UK in January 1951.
Photograph taken July, 1951

HMS *Glory*

HMS Glory
Firefly aircraft 812 Squadron over Sardinia, March 1950

HMS *Glory*, Bay of Biscay, 13 and 14 December 1950

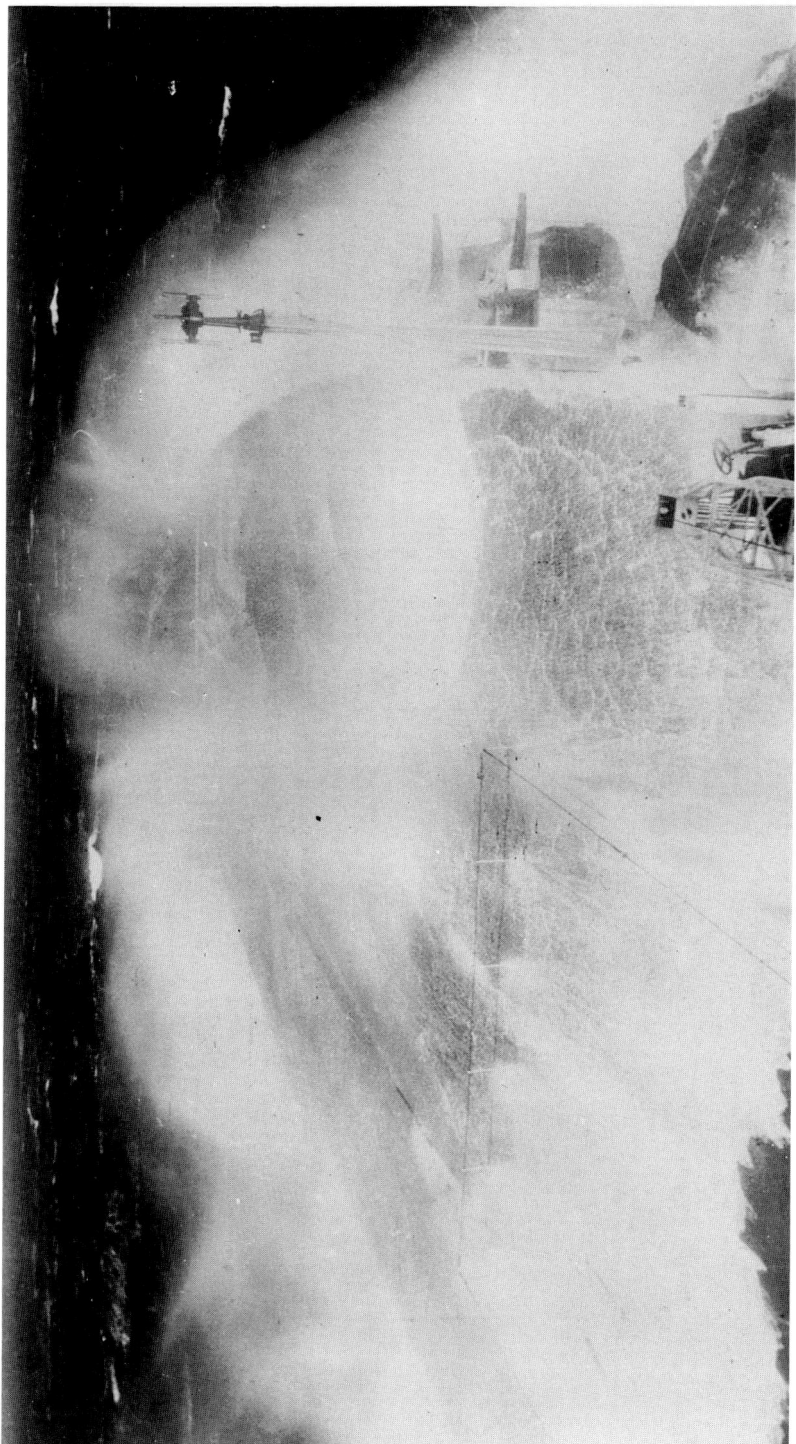

HMS *Glory*, Bay of Biscay, 13 and 14 December 1950

HMS *Glory*
HRH Princess Elizabeth inspecting Royal Marine Band, Grand Harbour
Malta, Wednesday 14 December 1949

HMS *Glory*
Firefly aircraft damaged when landing during night flying
(wooden propeller), 1949–50

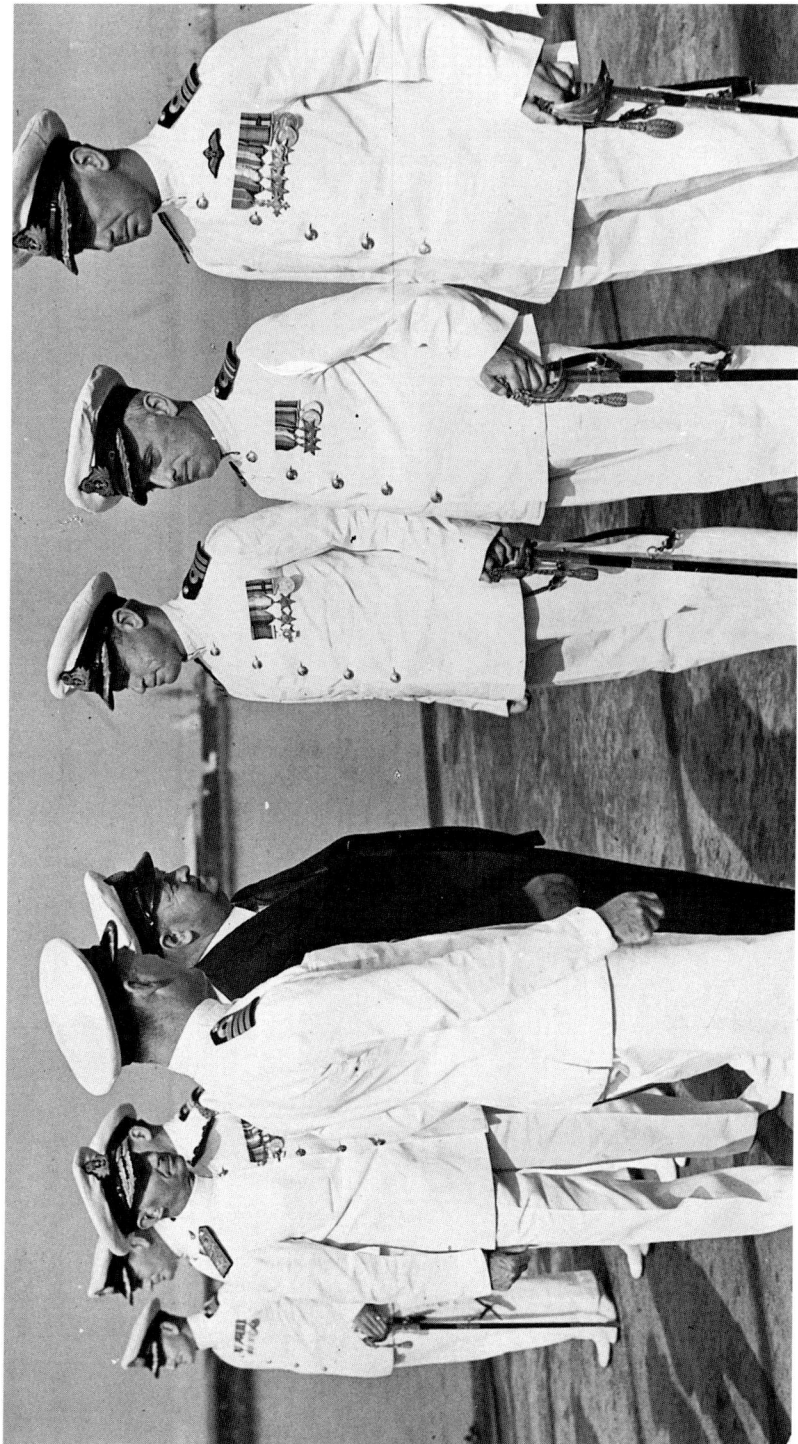

HMS Glory
Lord Hall inspects senior officers, Gibraltar, Monday 25 September 1950

HMS *Glory*

HMS *Glory*, Suez Canal, 1951

HMS *Glory*

HMS *Glory*
Fairey Firefly lands, Korea, 1951

HMS *Glory*

HMS *Glory*

HMAS *Bataan* and HMS *Glory* re-fuel at sea from RFA Wave Premier, Korean War, 1951

HMS *Glory*

ROLL OF **KOREA** HONOUR
1951 52 53

~1951~

LIEUTENANT (P) EDWARD PETER LANGDALE STEPHENSON. 28 APRIL
PILOT (3) STANLEY WILLIAM EDWIN FORD. 5 JUNE
LIEUTENANT (P) JOHN HARRY SHARP. 28 JUNE
AIRCREWMAN (1) GEORGE BERTRAM WELLS. 28 JUNE
LIEUTENANT (P) ROBERT WILLIAMS. 16 JULY
SUB LIEUTENANT (O) IAN ROBERTSON SHEPLEY. 16 JULY
COMMISSIONED PILOT TERENCE WILLIAM SPARKE. 18 JULY
SUB LIEUTENANT (O) RONALD GEORGE ALBERT DAVEY. 22 SEPT

~1952~

LIEUTENANT (P) RICHARD JAMES OVERTON. 15 MAR
LIEUTENANT (P) RICHARD NEVILL-JONES. 18 NOV
PETTY OFFICER (AIR) VICTOR COLMAN. 19 NOV

~1952~

LIEUTENANT (P) ALAN PHILIP DANIELS. 16 DE
AIRCREWMAN (1) ERNEST RAYMOND RIPLEY. 16 DE
LIEUTENANT (P) PETER GEORGE FOGDEN. 20 DE
LIEUTENANT (P) ROBERT EDWARD BARRETT. 25 DE

~1953~

SUB LIEUTENANT (P) BRIAN EDWARD RAYNER. 5 JA
SUB LIEUTENANT (P) JAMES MALCOLM SIMONDS. 5 JA
LIEUTENANT (P) CEDRIC ALEXANDER MACPHERSON 10 FE
SUB LIEUTENANT (P) RICHARD DEREK BRADLEY. 14 FE
LIEUTENANT (P) JOHN THOMAS McGREGOR. 25 AP
SUB LIEUTENANT (P) WALTER JOHN BUTLER KEATES. 25 AP

"LEST WE FORGET"

HMS *Glory*
Korean War Roll of Honour, 1951–53

HMS *Glory*
'Rum Issue', Leading Stores Assistant Bryan Green, checking the
Mess entitlement to 'grog'

133

HMS *Glory*

Sir Charles Lambe, CinC, Far East Fleet, visited *Glory* for a short time. He last came aboard in April 1953 when he had flown on an operational sortie over Korea. On the homeward voyage the first port of call was Colombo, the capital of Ceylon, after which began the long haul across the Indian Ocean. On Sunday 7 November, Divisions and a Remembrance Day service were held on the flight deck with two minutes silence being observed.

Whilst still a day out from Aden, *Glory* was hailed by the Royal Mail Steamer *Tantallon Castle*, who requested the services of *Glory*'s Medical Officer. At 1330 the seaboat was slipped and took the MO over to the *Tantallon Castle*. *Glory* steamed at five cables apart at a steady twelve knots for the next hour and a half when the MO came back aboard. During the short stay in Aden twenty three Royal Air Force personnel were given a conducted tour of the ship and afforded the usual hospitality afterwards. On Saturday 13 November the *Glory* left Aden at 0700 and headed for the Straits of Bab-el-Mandeb, the stretch of water connecting the Indian Ocean to the Red Sea.

This part of the voyage through the arm of sea which separates Arabia from Africa is fascinating. The numerous coral reefs and banks, together with the surrounding hills give a reddish tint to the water, which many believe is how the Red Sea got its name. Most of the reefs are near the coast, but the dangerous Daedalus reef is just below the surface and right in the middle. This and the myriad islands combined with the irregular currents of the Red Sea make navigation difficult and have been the cause of many ships foundering. At its northern end the Red Sea forks like a Y into two long narrow arms, the eastern arm is the Gulf of Aquaba, and the western arm forms the Gulf of Suez which *Glory* entered on Tuesday 23 November.

After an uneventful transit of the Canal, Sunday saw the ship between Port Said and Malta, with the morning service being held in the for'd lift well. Besides the usual stores and vehicles which were loaded at Malta for passage home, six polo ponies were also embarked on Thursday 25 November; they settled in very well and were cosseted during the journey. After a brief call at Gibraltar, *Glory* arrived in Portsmouth on Monday 6 December 1954 and secured alongside Pitch House Jetty.

The ship's next duty was far removed from the sweltering conditions of the Far East. On Wednesday 19 January 1955, *Glory*, in company with the frigate HMS *Urchin* anchored in Loch Eriboll on the north west coast of Sutherland, to take part in Operation Snowdrop, which had been quickly organised as the terrible blizzards struck most of the north of Scotland during the first weeks of the New Year.

From the flight deck naval and airforce helicopters took off in search of SOS signs in the snow. The previous evening after the Scottish news on the BBC, a request had been broadcast appealing to crofters and others requiring assistance to remember to light fires to give pilots an indication of wind direction, and to mark the ground with the code letter for the required assistance; 'D' for doctor, 'F' for food, 'C' for cattle fodder. Relief work was in many cases being hampered as this had not been done and pilots were therefore unable to say whether they had delivered supplies to the people who required them most urgently. People were also asked to put oil or material onto their fires as the chimney smoke from peat fires was so light in colour that aircraft in the vicinity might miss the signals. *Glory* had sailed from the Clyde with two thousand tons of fuel for the helicopters and her ship's company made up food parcels and baked loaves of bread by the thousands. An indication of the severity of the weather in the Burgh of Inverness was provided by the Town Road Department workmen clearing away snow from the centre of the town, when in Church Street they came upon the body of a full grown spaniel which had evidently been frozen to death.

Glory returned to Rosyth after her mercy mission and work commenced on refitting. It was during this time that the Admiralty, according to newspaper reports, was considering the future of the *Glory*, *Theseus*, *Ocean* and *Unicorn* as it was thought they were 'surplus to requirements'.

Glory spent the rest of 1955 in Rosyth reduced to care and maintenance. It was during this time that *Glory* became known as the 'ghost ship'. A dockyard worker employed in refitting operations claimed to have seen the ghost when he entered an empty cabin at about 0730. He declared he saw a man in tropical flying kit. When the figure disappeared, the workman fled the ship.

Rumour quickly spread that the 'ghost' was the apparition of a pilot killed when making a crash landing aboard the *Glory* after an action during the Korean War. Despite denials by the officer-in-charge that no pilot had been killed in a crash landing on board *Glory*, the ship continued to be known as a 'ghost ship'.

On Tuesday 10 April 1956 at 1845, Captain T.N. Masterman RN arrived to take command of *Glory* and work began on preparing the ship for sea. On Thursday 10 May she left her berth at Rosyth, and proceeded to the anchorage below the Forth Bridge, securing to number 52 buoy. The following day *Glory* left Scotland and made her way to Plymouth, stopping for the day in Portland Harbour on Sunday. The ship secured to number nine buoy in the Hamoaze on Monday, and the next few weeks were spent in the West Country. *Glory* returned to Rosyth Dockyard on Friday 15 June 1956, and one week later on Friday 22 June, the ship's log was stamped 'Paid Off' and signed by Captain T.N. Masterman OBE RN. Sometime in 1957, all preservation work on *Glory* was suspended. In April 1958 it was reported by the Belfast Telegraph, that *Glory* would be broken up within the 'next few weeks'.

In November of that year her future was once again being reconsidered, it was thought she might be converted for troop carrying duties. Investigations went on for some weeks but in the end came to nothing. In April 1961, the *Glory*'s days of fame were drawing to a close, she had been on the list for 'disposal' for some time, and eventually the Admiralty reached their decision. On Wednesday 23 August 1961, HMS *Glory* was towed by the tugs *Impetus*, *Handmaid* and *Energy*, with smoke belching from their funnels, the short distance down the river to Thomas W. Ward Ltd, the breakers yard at Inverkeithing. After *Glory* had been secured for the last time, the manager of the breakers yard Mr W. Gray, on being told of the *Glory*'s 'ghost', said 'There won't be any place left for the ghost after we have finished with her.'

Eight months later and the burners had completed their work. The *Glory*'s name however, is carried on. There is a small but thriving *Glory* Association, which continues to meet twice a year. One of the members, Mr David Wharton and his wife June, have been to Grenfell, New South Wales, where they were able to dedicate a seat in Taylor Park to the Australian people for their kindness and hospitality to the crew of HMS *Glory* in 1945 and 1951. On Sunday 20 September 1992, a standard which has been kindly donated to the HMS *Glory* Association by Mr Bert Roach, was dedicated in the St Paul's Naval Church, HM Naval Base, Portland, by the Chaplain. Meanwhile, the Horsham Sea Cadet Corps proudly wears the cap tally, *T.S. Glory*.

Appendices

1. Ship's Captains.
2. Newspaper Article Thursday 26 July 1945.
3. Crossing line certificate Philip Lister.
4. HMS *Glory* at Sydney 1945.
5. Commander's Orders 5 September 1945.
6. Daily Orders Thursday 6 September 1945.
7. Japanese Surrender General Programme.
8. Flying Programme Thursday 6 September 1945.
9. Directive by Lieut-Gen V.A.H. Sturdee to General Imamura.
10. Instrument of Surrender Document.
11. Signal regarding repatriation of POWs.
12. Glory News Vol II No 3, 28 February 1946.
13. First Commission Ports of Call.
14. First Commission Roll of Honour.
15. Second Commission Ports of Call (Spring Cruise).
16. Second Commission Ports of Call (Summer Cruise).
17. Commander's Temporary Memorandum No 117.
18. First, Second and Third Tours of Korea, 1951–53.
19. Operations in Korean Waters (First Tour).
20. Operations in Korean Waters (Second Tour).
21. Press Release.
22. Operations in Korean Waters (Third Tour).
23. Flying Programme Sunday 9 September 1951.
24. Flying Programme Thursday 14 May 1953.
25. Flying Programme Sunday 5 April 1953 (Record Breaker).
26. Signal regarding 123 operational sorties.
27. Flying Programme 29 April 1952.
28. Stoker David Hyam's Identity Certificate.
29. Signal from FO2FES to Admiralty.

HMS *Glory*

(Per Concordiam Gloria)
Light Fleet Aircraft Carrier
Ship's Captains
1945–1956

Captain	A.W.	Wass-Buzzard	RN
Captain	W.T.	Couchman	RN
Captain	H.A.	Traill	RN
Captain	E.H.	Shattock	RN
Captain	K.S.	Colquhoun	RN
Captain	T.A.K.	Maunsell	RN
A/Captain	D.E.	Bromley-Martin	RN
Captain	E.D.G.	Lewin	RN
Captain	R.T.	White	RN
Captain	H.W.	Sims-Williams	RN
Captain	T.N.	Masterman	RN

Glorious 1 June 1794   Martinique 1809

Guadaloupe 1810   Dardanelles 1915

Korea 1951–53

Newspaper Article Thursday 26 July 1945
Washington

Senator Thomas Hart, formerly the Admiral commanding the US Pacific Fleet, now a Republican Senator, speaking in the debate on the United Nations Charter said, 'it is known that a detachment of the British Navy is fighting with our forces in the north west Pacific, Admiral Nimitz does not need those forces because by the time they joined him he had already beaten the Japanese Navy at sea and in the air. By the date of the invasion of Normandy allied naval forces had fairly well won their part of the Atlantic war.

'If three months ago the British Navy had brought their amphibious power into action from the Indian ocean an extremely effective contribution to the Pacific war would have resulted. Why was the drive from the Indian ocean not started some time ago? To my mind the reason the British did not drive into the Netherlands East Indies and through the Malay barrier was simply because they felt they could not.'

PROCLAMATION

Be It Known that Philip J R Lister

did on this Fourth day of August 1945 being embarked in His Majesty's Ship GLORY, cross the Equator at Longitude 87° 05' East and was duly presented to our Royal Court.

By Virtue Whereof, I, Neptunus Rex, Ruler of the Raging Main, do declare him to be truly initiated into the mysteries of the Tropical Deeps.

Neptunus R.
RULER OF THE MIGHTY DEEP.

GLORY

HMS *Glory* at Sydney 1945

Not it seems an old forgotten fable:
The snow goose descending on the still lagoon,
And trees of summer flowering ice and fire
And the sun coming up on the blue mountains.

But I remember, I remember Sydney,
Our bows scissoring the green cloth of the sea,
Prefaced by plunging dolphins we approached her:
The land of the kookaburra and the eucalyptus tree.

The harbour bridge, suddenly sketched by Whistler,
Appeared gently on a horizon of smudges and pearls,
And the sun came up behind us
With a banging of drums from the Solomons.

O! I shall never forget you on that crystal morning!
Your immense harbour, your smother of deep green trees,
The skyscrapers, waterfront shacks, parks and radio-towers,
And the tiny pilot boat, the *Captain Cook*,
Steaming to meet us:
Our gallery deck fringed with the pale curious faces of sailors
Off the morning watch.

O like maidens preparing for the court ball
We pressed our number-one suits,
Borrowing electric irons and starching prim white collars,
And stepped forth into the golden light
With Australian pound notes in our pockets.

O there is no music
Like the music of the Royal Marine bugler
Sounding off 'Liberty men'.
And there is no thrill
Like stepping ashore in a new country
With a clean shirt and with pound notes in your pocket.

O Sydney, how can I celebrate you
Sitting here in Cornwall like an old maid
With a bookful of notes and old letters?

I remember the circular bar in Castlereagh Street
And the crowds of friendly Aussies with accents like tin-openers,
Fighting for schooners of onion beer.
I remember Janie, magnificent, with red hair,
Dressed in black, with violets on her reliable bosom,
Remembering a hundred names and handling the beer engines
With the grace and skill of ten boxers.

O Janie, have the races at Melbourne seen you this year?
And do matelots, blushing, still bring you flowers?
Across three continents: across monsoon, desert, jungle, city,
Across flights of rare birds in burning Africa,
Across crowds of murderous pilgrims struggling grimly to Mecca,
Across silver assaults of flying-fish in the Arabian Sea,
I salute you and your city.

I remember the deep canyons of streets, the great shafts of sunlight
Striking on fruit-shop, flower-shop, tram and bookstall,
The disappearing cry of the Underground Railway,
The films: *Alexander Nevsky* and *Salome*,
The plays: *Macbeth* and *Noah* in North Sydney,
And travelling there, across the fantastic bridge,
Our ship, the *Glory*, a lighted beetle,
A brilliant sarcophagus far below
On the waterfront at Woolloomoolloo.

O yes I remember Woolloomoolloo,
The slums with wrought-iron balconies
Upon which one expected to find, asleep in a deck-chair,
Asleep in the golden sun, fat, grotesque and belching:
Captain Cook.

The Chinese laundries, the yellow children in plum-coloured brocade,
The way they fried the eggs, the oysters and champagne.
I remember Daphne and Lily, the black-market gin,
And crawling back to the docks as the dawn
Cracked on my head.

145

O the museum with the gigantic, terrifying kangaroo,
Who lived, as huge as a fairy story,
Only ten thousand years ago.

O the sheepskin coats, the woollen ties,
And our wanderings in David Jones' store
Among a rubble of silk stockings and tins of fruit salad.
The books I bought at Angus & Robertson's bookshop,
Sir Osbert Sitwell, and Q (to remind me of home).

I remember the ships and ferries at Circular Quay,
And the tram ride to Botany Bay,
So magnificently like the postage stamp
I bought as a child.
I remember the enormous jail at La Perouse,
The warders on the walls with their rifles.
I remember the Zoo at Taronga Park,
The basking shark I gazed down at in terror,
And the shoes I wore out walking, walking.

And so I celebrate this southern city
To which I shall never return.
I celebrate her fondly, as an old lover,
And I celebrate the names of my companions:

George Swayne, Ron Brunt, Joney,
Tug Wilson, Jan Love, Reg Gilmore,
Pony Moor, Derby Kelly, Mac.

Where are they now?

Now it seems an old forgotten fable:
The snow goose descending on the still lagoon,
The trees of summer flowering ice and fire
And the sun coming up on the Blue Mountains.

                                        Charles Causley

146

Commanders Orders
HMS *Glory*
5 September 1945

Japanese Surrender

1. *Sequence of Events*
   (a) Delegates arrive by boat (2nd MB) at Starboard after ladder.
   (b) Led straight aft starboard side to Quarter deck.
   (c) All arms to be surrendered and tallied (MAA and RPO's).
   (d) Delegates led up starboard passage to 'C' Hangar and on to lift.
   (e) Up to Flight Deck.
   (f) Walk forward on Flight Deck between Divisions to GOC.
   (g) Surrender signed and proclamation read.
   (h) GOC and staff leave via Island.
   (i) Japanese return to Quarter deck via after lift and 'C' Hangar.
   NOTE: If wet the surrender will take place in the hangar but the sequence of events will be the same.

2. *Accommodation for meals*
GOC	Captain's Cabin
C in C	Commander's Cabin
Japanese Staff	Intelligence Room

3. *Officers*
Attending on GOC	Sub-Lieut Mason
In Intelligence Room	Sub-Lieut

4. *Sentries*
   All sentries will be armed with Lanchester Carbines.
Outside Commander's Cabin	Marine
Outside Intelligence Room (meals only)	Marine
Quarter deck Lobby	Seaman
Passage leading to Quarter deck each side	Seaman

5. *Messengers*
Captain's Lobby	Seaman
Quarter deck Lobby	Seaman

6. *Dress*
Officers	No 10
Dutymen	No 6 or 5 and collars
Cruising watch and Divisions	No 6 or 5 and collars

   Aircraft handling party and squadron ratings on duty working rig – to be kept out of sight.

(Signed) J.N. Hicks
Commander

Daily Orders
Thursday 6 September 1945

1st Duty Lieutenant Commander	Lieutenant Commander Thomas
2nd Duty Lieutenant Commander	Lieutenant Commander Creed
1st EOOD	Lieutenant (E) Dawe
2nd EOOD	Mr Brading
Air OOD	Lieutenant Killick
Air EOOD	Sub Lieutenant Evans
Duty Medical Officer	Surgeon Lieutenant Evans
Duty Warrant Officer	Mr Drew
Duty Supply Officer	Lieutenant Dear
Duty Watch	RED
Duty Boat	2nd Motor Boat

ROUTINE
Daily Sea Routine – Hands required in the afternoon dress of the day:
No 6's

0700	Forenoon watchmen clean in Rig of the Day during breakfast.
1000	Hands to clean into No 6's (or 5's with collar).
1010 (about)	Guns crews lower 2nd Motor Boat and Starboard after ladder.
1025 (about)	Hands to Divisions on the Flight Deck. Divisions to be six deep and closed well aft. Unattached officers forward.
1030 (about)	Japanese delegates arrive. Guns crews hoist 2nd Motor Boat and Starboard after ladder.
	After divisions hands clean into Working Rig.
1430 (about)	Lower 2nd Motor Boat and Starboard after ladder for departure of Japanese delegates.

Flying programme issued separately.

No training.

NOTE: Special orders for the Japanese surrender are on Ship's Company Notice Board.

(Signed) J.N. Hicks
Commander

HMS *Glory*
Wednesday, 5 September 1945

Rabaul Surrender
Programme – Thursday 6 September 1945

A. GENERAL PROGRAMME

0600 Vendetta R/V Hart in approximate position 4° 50″ South 152° 35″ East. Proceed in company to Kabanga Bay.

0800 Vendetta and Hart arrive Kabanga Bay. Latter embarks Japanese and returns to *Glory*. Former proceeds to Rabaul.
NOTE: Japanese party is maximum of 12.

1030 (approx) Hart R/V *Glory* and transfers Japanese.

1045 to 1115 (approx) Surrender ceremony on board *Glory*.
NOTE: Flight Deck closed from 1000 to 1130.

1100 Air demonstration over Rabaul.

1115 to 1430 (approx) Conference with Japanese staff.

1445 Transfer Japanese party (and any mail) to Hart, allied officers (and any mail) to *Amethyst*.

1500 Hart leaves for Kabanga Bay. *Amethyst* for Jacquinot Bay.
NOTE: Flight deck closed for flying from 1430 to 1600.

1630 Hart arrives Kabanga Bay, lands Japanese and returns to *Glory*.

1800 (approx) Hart rejoins *Glory*, and is detached as necessary to Jacquinot Bay. As required *Glory* proceeds to Lae with GOC and Staff.

B AIR COMMITMENTS

1. *Glory*:
   (a) Own CAP (all day)
   (b) Demonstration over Rabaul (1050 to 1100)
   (c) CAP for Hart to and from Kabanga Bay (1600 to 1715)

2. *RNZAF Jacquinot Bay*:
   (a) CAP for Hart and Vendetta to and from Kabanga Bay. (0700 to 1030).
   (b) CAP for Vendetta and M/S off Rabaul (1030 to 1230).
   (c) Demonstration over Rabaul (1100).

(Signed) D.G. Goodwin
Commander (Operations)

Air Office
5 September 1945

FLYING PROGRAMME
– Thursday 6 September 1945

0830   Land on first CAP of 4 Corsairs from Jacquinot Bay.
            Range 6 Barracuda and 8 Corsairs for flying off.
            (No long range tanks.)
            Strike down 4 CAP Corsairs.

0940   Fly off 8 Corsairs and 6 Barracuda for demonstration over
            Rabaul.

1050–1100   Own aircraft demonstrate over Rabaul.

1100   Second CAP from Jacquinot Bay returns to base.

1135   Land on 6 Barracuda and 8 Corsairs from demonstration.
            Strike 6 Corsairs down into A hangar.

1330   Land on third CAP of 4 Corsairs and 1 Barracuda from
            Jacquinot Bay.
            Strike 5 Barracuda down into B hangar and 1 Barracuda into
            C hangar.

1600   Land on 4th CAP of 4 Corsairs from Jacquinot Bay.
            Fly off 2 Corsairs CAP for sloop.

1715   Land on 2 Corsairs.

(Signed) H.P. Sears
Commander (Flying)

DIRECTIVE
by
LIEUT-GEN V.A.H. STURDEE CB CBE DSO
General Officer Commanding First Australian Army
to
GENERAL IMAMURA
6 September 1945

The Japanese empire has been defeated in battle.

Your Emperor and your government have surrendered unconditionally to the allied powers.

The Japanese imperial general staff have issued orders that Japanese forces everywhere will lay down their arms and submit to the orders and instructions of the allied powers.

I understand from your signal of 3 September 1945 that you have received such orders from Tokio. My representative handed to your envoy off Rabaul on 4 September a copy of such orders for transmission to you.

You are to comply forthwith with this order and any subsequent orders and instructions that may be issued by me or my representatives.

You are hereby appointed Commander of all Japanese army and navy forces and all other Japanese in the area under my command.

You will be the sole Japanese authority in the area to whom I will issue official orders, directions and instructions.

You are to issue orders immediately to your commanders in New Britain, New Ireland, Bougainville and New Guinea to surrender to my commanders in these areas, and for your commanders to comply with General Order No 1 Army and Navy. When this has been done, your command and jurisdiction will be confined to the Rabaul area and New Ireland.

My representatives, under whose detailed orders the occupation of the respective areas and the confinement of your troops will be carried out, are:

NEW BRITAIN AND NEW IRELAND
Maj-Gen K.W. Eather
GOC 11 Aust Div

BOUGAINVILLE
Lieut-Gen S.G. Savige
GOC 2 Aust Corps

NEW GUINEA
Maj-Gen H.C.H. Robertson
GOC 6 Aust Div

Your forces will immediately and rigidly comply with the orders to disarm. Any Japanese, other than authorised police or guards, discovered in possession of arms or explosives of any nature are liable to be shot on sight without enquiry or trial.

Your forces and Japanese civilians will be confined in such places and under such conditions as I or my representatives may order.

All Japanese will be treated with firmness, justice and humanity.

They will construct their own accommodation and areas of confinement, and perform such other work as may from time to time be ordered. They will farm the land so as to produce the maximum food for their own sustenance and that of the Australian forces, allied prisoners of war and local natives.

The local production of the maximum quantity of food is essential to the well being of the Japanese confined within all areas, as the shipping available and quantity of food that can be imported is definitely limited.

I have been duly authorised by the Commander-in-Chief Australian Military Forces to accept your surrender and that of the Japanese army and navy forces and other Japanese nationals under your command.

I now order that you sign the instrument of surrender.

HMS *Glory*   Lieut-Gen
6 September 1945   GOC First Australian Army

ADMIRALTY, NSHQ OTTAWA, CTF 111
VA (Q) BPF, COMPHILSEAFRON, C-IN-C PAC,
BAD WASHINGTON (R) COMINCH, AC1
IMPLACABLE, FORMIDABLE, GLORY
C-IN-C AF PAC, SCAP, SACSEA,
C-IN-C EI, CTF 57, ACMB, C-IN-C HONG KONG

From: C-in-C BPF 200815Z

Operational Priority to BAD, C-IN-C AFPAC, SCAP, COMPHILSEAFRON

PRIORITY TO REMAINDER

Unclassified

My 200512 and 200522 not to or needed by all. Unless otherwise directed by C-in-C AFPAC following are my intentions for employment of CV's *Implacable* and *Formidable* and CVL *Glory* to lift processed RAPWI from Manila.

2. Ships will lift RAPWI whose final destination is United Kingdom or Canada. Ships will proceed to Vancouver or Seattle as directed by BAD, calling at Pearl for fuel subject to concurrence of C-in-C PAC.

3. Approximate dates of arrival on West Coast are as follows:

Implacable	11 October
Formidable	17 October
Glory	23 October

4. Numbers of RAPWI carried as follows:

Implacable	275 sick and 1,475 fit or	2,125 fit.
Formidable	250 sick and 900 fit or	1,400 fit.
Glory	260 sick and 870 fit or	1,300 fit.

Ships are to signal BAD repeated C-in-C BPF number embarked before sailing from Manila.

5. BAD is requested to make the necessary arrangements as in my 080845Z, para 5, as confirmed by his 111534 neither to all.

6. BAD is requested to arrange for fuel on the West Coast. Ships are to signal their requirements on leaving Pearl.

7. As ships may require to make a second trip, it is desired that they be stored to capacity before leaving America. BAD is requested to arrange that if possible, ships are to report earliest practicable to state:

(a) Their re-storing requirements.
(b) Any special requirements of RAPWI on landing (such as clothing).

8. COMPHILSEAFRON pass to SBNLO PHIL and COMGENAFWESPAC. BAD WASHINGTON pass to BAS NSHQ OTTAWA pass to BATM.

9. RDO Honolulu pass to TDO Tokio for C in CAFPAC and RDO Yokohama from SCAP also to C-in-C PAC.

=200815z

HAND.   P/L   TOR 210930/September 1945.   PL.

Glory News

Vol II No 3   28 February 46

EGYPTIAN CRISIS GROWS AGAIN

Egyptian students say that unless British troops are evacuated from main Egyptian cities they would lead a nationwide uprising on Friday 1 March. The situation seems to be worsening again daily. A student statement says 'Students demand immediate action! If none appears by Friday, underground and sabotage units will be advised to attack the British'. This action is commented as taking the shape of a 'Holy War' on the part of the students. A further statement declares Monday is to be a day of mourning for Egyptians killed in the recent disturbances.

The British attitude remains firm. Viscount Addison told the House of Lords that the Government cannot acquit the Egyptian Government of responsibility for the recent disorders. He again emphasised the Government's determination to replace the Treaty of Alliance with a fresh treaty negotiated between equals.

A correspondent in Cairo announced the British barracks and military headquarters in Cairo were attacked before the incident named by Sidky Pasha as the original cause of the rioting in which the drivers of British military vehicles were accused of angering the crowd.

RUSSIA–CHINA DISPUTE

Reports from Manchuria seem to suggest that the traditional Russian motto of 'act first talk afterwards' is being implemented again in Manchuria.

Moscow has indicated that machinery is being moved from former Japanese factories to Russia and claims that this was agreed at the meeting of the Big Three.

Mr Byrnes, the US Secretary of State, denies knowledge of any such agreement ...

The Russians state that the delay in moving troops after the agreed date of evacuation of 1 February was due to the fact that remnants of the Japanese forces have been making widespread bandit raids. The evacuation has now commenced and Russians are being replaced by Chinese troops.

PRIME MINISTER ON INDIAN MUTINY

Mr Attlee yesterday announced that a full enquiry would be held into RIN by a Government of India Committee Association with members of a central legislative assembly. Also there would be courts of enquiry held by RIN.

In Bombay, Congress Party leaders condemned recent disorders there. One of them, Nehru, called for an inquiry into police and military action. There was more trouble in Madras area yesterday. Police had to fire on crowds that held up an express train outside the city.

Late Stop Press

Spain
Frontier despatches reaching Paris a few hours ago said 'Spain closed the French-Spain frontier this morning' nearly 48 hours before the announced closing time by the French.

AROUND AND ABOUT

Bali
Six months after the end of the war in the Pacific the Allies accepted the formal surrender of the Japanese on the island of Bali on Sunday. Famed for its beautiful girls, this island paradise has been under Japanese administration until the surrender took place on board HM frigate 'Loch Esk'.

Tokio
The Japanese Government announced yesterday its decision to postpone the general elections for 31 March to 10 April. The reason given was that more time was needed to consider the eligibility of candidates under the purge directed from Allied HQ.

London
Government reports on rehousing show that the total of permanent houses completed in England, Scotland and Wales up to the end of January was 3,468 and temporary bungalows totalled over 12,000.

Nuremburg
The International Court yesterday rejected a plea made by the former Nazi foreign minister, von Ribbentrop, that Mr Churchill and several pre-war French leaders be summoned to witness on his behalf. A list of

38 names was presented by him, 22 of which were rejected. Lists of witnesses submitted by Goering and Keitel were also trimmed by the court.

The Soviet prosecutor accused Goering and Keitel of being responsible for the execution of 50 RAF officers after an unsuccessful attempt to escape from a Silesian prison camp a year ago.

Palestine
A search is going on in Palestine for bands of terrorists who yesterday attacked five RAF stations and destroyed 12 aircraft.

Washington
The House of Representatives on Tuesday approved and sent to the White House for signature a Bill regulating the disposal of $17,000,000,000 worth of surplus war built merchant (Liberty) ships.

Sale prices are fixed between the limits of not more than 7/8 and not less than 31½ per cent of the original cost.

New Zealand
The Minister of Supply, Mr Sullivan, stated that stocks of petrol in the Dominion are sufficient only for one month. His request to London for increased shipments has been refused on the grounds that supplies from the sterling area could not be increased and no dollars could be spared for the purchase of petrol from America.

The private ration in New Zealand ranges from four to eight gallons per month.

ANNOUNCEMENTS

Any notices on entertainments, sports, lost and found notices, or any other contributions to 'News' should be handed into the Press Office between 1800 and 1900. Press Office is temporarily situated in the Army Liaison Office in the Captains flat.

TELEGRAMS

When sending telegrams please bring the exact amount – changing is difficult.

AUSSIE TOPICS

Sydney
Australian dockers and wharf labourers announced yesterday that they would load one Dutch ship with food and medical supplies for Java. This reversal of their decision made several months ago is a trial and if the Commonwealth Government can satisfy them that no arms or military supplies find their way abroad, full shipments will be resumed. New Zealand dockers took a similar decision last week.

First Jap ship to visit Sydney for nearly five years arrived there yesterday to repatriate nearly 3,000 Jap prisoners of war. She was the *Diki Maru* and was escorted by two British submarines.

Canberra
Prime Minister Chiffley stated yesterday that Australia has been granted a portion of the store of Japanese raw silk. The amount would not be more than 520 tons. No wheat would be sent to Japan in exchange as was first requested.

Mr Caldwell, Information Minister, said there were good prospects of bringing wives, children and fiancées of Australian servicemen from Britain to Australia in two large ships now being refitted if *Queen Mary* or *Queen Elizabeth* were not available. Australian Minister in London, Mr Beasley, had succeeded in doubling large quota of shipping for Australia.

The 13 members of the Australian Cricket team arrived at Auckland on Tuesday in an RNZAF Catalina. They will tour the four main centres and finish off with a test in Wellington.

WAR DATA

Due to the rather prolific propaganda of other nations and the variances in time when war-effort statements have been announced, it has been difficult to get a clear picture of what the British Empire has done and suffered throughout the war years.

Glory News will publish a few facts each day and try and make this clearer and state exactly what the Empire's burden has been in comparison with the United States.

RN AND USN LOSSES IN SHIPS
Light coastal craft and landing craft not included

Capital Ships	5 (RN)	2 (US)
Aircraft Carriers	8 (RN)	11 (US)
Cruisers	26 (RN)	10 (US)
Destroyers	128 (RN)	61 (US)
Submarines	77 (RN)	52 (US)
Minesweepers	51 (RN)	24 (US)
Trawlers	240 (RN)	
Corvettes	20 (RN)	
Frigates	11 (RN)	11 (US)
Sloops	10 (RN)	

PERSONNEL CASUALTIES

Overall Empire Casualties in the War	1,233,796 (Dead 336,772)
Overall UK Casualties in the War	750,338 (Dead 233,042)
Overall RN Casualties in the War	66,000 (Dead 49,305)
US Losses Army	922,583 (Dead 207,750)
US Losses Navy	147,570 (Dead 55,400)

UK Total of killed was .007 of the population
US Total of killed was .0018 of the population

First Commission Ports of Call

Wednesday 21 February 1945	Belfast, N. Ireland
Monday 14 May 1945	Glasgow, Scotland
Monday 21 May 1945	Marsaloxx, Malta
Thursday 24 May 1945	Alexandria, Egypt
Thursday 12 July 1945	Trincomalee, Ceylon
Friday 10 August 1945	Fremantle, Australia
Thursday 16 August 1945	Sydney, Australia
Thursday 6 September 1945	Rabaul, New Britain
Tuesday 2 October 1945	Manus, Bismark Islands
Saturday 6 October 1945	Leyte, Philippine Islands
Tuesday 9 October 1045	Manila, Philippine Islands
Saturday 20 October 1945	Pearl Harbour, Oahu Island
Friday 26 October 1945	Esquimalt, Canada
Saturday 27 October 1945	Vancouver, Canada
Wednesday 21 November 1945	Hong Kong, China
Saturday 24 November 1945	Manila, Philippine Islands
Wednesday 28 November 1945	Balikpapan, Borneo
Saturday 1 December 1945	Tarakan, Borneo
Friday 7 December 1945	Manus, Bismark Islands
Wednesday 12 December 1945	Sydney, Australia
Wednesday 23 January 1946	Melbourne, Australia
Thursday 14 February 1946	Sydney, Australia
Saturday 2 March 1946	Auckland, New Zealand
Saturday 23 March 1946	Kure, Japan
Friday 12 April 1946	Sydney, Australia
Saturday 15 June 1946	Newcastle, Australia
Thursday 20 June 1946	Sydney, Australia
Thursday 27 June 1946	Adelaide, Australia
Monday 15 July 1946	Trincomalee, Ceylon
Tuesday 1 October 1946	Hong Kong, China
Monday 18 November 1946	Singapore, Malaya
Thursday 19 December 1946	Hong Kong, China
Saturday 8 March 1947	Bombay, India
Tuesday 25 March 1947	Trincomalee, Ceylon
Saturday 17 May 1947	Singapore, Malaya
Thursday 3 July 1947	Adelaide, Australia
Friday 11 July 1947	Melbourne, Australia
Thursday 24 July 1947	Sydney, Australia
Friday 8 August 1947	Brisbane, Australia
Saturday 20 August 1947	Singapore, Malaya
Sunday 7 September 1947	Trincomalee, Ceylon
Wednesday 17 September 1947	Aden, South Arabia
Friday 26 September 1947	Valletta, Malta
Thursday 2 October 1947	Gibraltar, Spain
Monday 6 October 1947	Plymouth, England
Tuesday 7 October 1947	Glasgow, Scotland
Tuesday 14 October 1947	Plymouth, England

First Commission

ROLL OF HONOUR

Sub. Lieutenant Snape RN	15 April 1945
Sub. Lieutenant Pickles RN	15 April 1945
Leading Airman Ryan	15 April 1945
Able Seaman Emerson	18 April 1945
Able Seaman Thompson	1 June 1945
Sapper William Owens	14 October 1945
Sub. Lieutenant Sanderson RN	22 January 1946
Corsair Pilot	7 February 1946
Steward Cross	7 February 1946
Leading Airman Berry	8 February 1946
Sub. Lieutenant Lawson	18 June 1946
Naval Airman Adams	27 September 1946
Lieutenant MacKinnon RN	5 March 1947
Lieutenant Commander Hamilton-Bates RN	7 March 1947
Lieutenant Mayne RN	7 March 1947
Lieutenant Commander Thurston RN	29 April 1947
LAM (E) Walmsley	2 June 1947
Naval Airman Saddler	23 July 1947

HMS *GLORY* COMMANDING OFFICERS 1945–1947

Captain Sir Anthony Wass Buzzard, OBE, DSO, Royal Navy
Captain W.T. Couchman, OBE, DSO, Royal Navy

Mediterranean Fleet Spring Cruise 1950

Thursday 26 January 1950	Naples, Italy
Friday 3 February 1950	Tripoli, Libya
Tuesday 7 February 1950	Valletta, Malta
Thursday 2 March 1950	Palmas Bay, Sardinia
Wednesday 8 March 1950	Golfe Juan, France
Thursday 16 March 1950	Algiers, Algeria
Wednesday 22 March 1950	Gibraltar, Spain
Friday 31 March 1950	Valletta, Malta

Mediterranean Fleet Summer Cruise 1950

Wednesday 14 June 1950	Corfu, Greece
Friday 23 June 1950	Skiathos, Greece
Saturday 1 July 1950	Piraeus, Greece
Saturday 8 July 1950	Marmararice, Turkey
Friday 14 July 1950	Larnaca, Cyprus
Saturday 22 July 1950	Alexandria, Egypt
Wednesday 2 August 1950	Valletta, Malta
Friday 8 September 1950	Marseille, France
Saturday 16 September 1950	Tangier, Morocco
Thursday 21 September 1950	Gibraltar, Spain
Monday 2 October 1950	Valletta, Malta

HMS *GLORY*   COMMANDER'S OFFICE
                                          15 September 1950

COMMANDER'S TEMPORARY MEMORANDUM No 117
TANGIER

Arrive at 0900 Saturday 16 September.

Depart Wednesday 19 September.

We shall anchor about a mile from the landing place. An MFV from *Gibralter will be attached to the ship for liberty trips etc. HMS *Saintes* will be in company.

The ship will be open to visitors from 1400 to 1830 on Monday 18 September.

Leave will be granted to one watch until 0030, Chief and Petty Officers 0100.

Transport Buses run to the suburbs of Tangier and are very crowded. They are mostly Spanish owned and prices of journeys vary with the rate of exchange of the *pesata. Taxis are fairly plentiful, but expensive. It is wise to fix the price of a journey before starting.

Entertainments and Recreation Excellent swimming beaches line the Bay of Tangiers and the Atlantic coast of Cape Spartel. Both theatres and cinemas are poor. Plays are in Spanish and films are old.

Restaurants and bars are numerous and of all types. The Hotel Bretagne, on the front, is recommended for food. The quality of liquor is good and there are no special brands to avoid.

There are plenty of guides. Those authorised by the British Legation carry a stamped card and are fairly reliable.

Shopping is cheaper than in *Gibralter. The usual Mediterranean seaport *deseases – including VD – are prevalent.

                                      P.N. BUCKLEY
                                      COMMANDER

* Spellings as on original document.

GLORIOUS 1st JUNE 1794   MARTINIQUE 1809
GUADALOUPE 1810   DARDANELLES 1915
KOREA 1951–2–3

FIRST TOUR

Thursday 25 January 1951	Devonport Departed
Friday 2 February 1951	Malta Arrived
Wednesday 21 March 1951	Port Said Arrived
Monday 26 March 1951	Aden Arrived
Friday 6 April 1951	Singapore Arrived
Tuesday 10 April 1951	Hong Kong Arrived
Monday 23 April 1951	Sasebo Arrived
Sunday 30 September 1951	Kure Departed
Wednesday 3 October 1951	Hong Kong Arrived
Tuesday 9 October 1951	Singapore Arrived
Wednesday 17 October 1951	Fremantle Arrived
Wednesday 24 October 1951	Sydney Arrived

SECOND TOUR

Monday 7 January 1952	Sydney Departed
Thursday 17 January 1952	Fremantle Arrived
Wednesday 23 January 1952	Singapore Arrived
Wednesday 30 January 1952	Hong Kong Arrived
Tuesday 5 February 1952	Sasebo Arrived
Thursday 1 May 1952	Sasebo Departed
Saturday 3 May 1952	Hong Kong Arrived
Sunday 11 May 1952	Singapore Arrived
Tuesday 20 May 1952	Aden Arrived
Friday 23 May 1952	Port Said Arrived
Monday 26 May 1952	Malta Arrived
Wednesday 23 July 1952	Istanbul Arrived
Wednesday 30 July 1952	Cyprus Arrived
Friday 15 August 1952	Malta Arrived
Monday 15 September 1952	Barcelona Arrived

THIRD TOUR

Thursday October 1952	Malta Departed
Sunday 12 October 1952	Port Said Arrived
Friday 17 October 1952	Aden Arrived
Tuesday 28 October 1952	Singapore Arrived
Tuesday 4 November 1952	Hong Kong Arrived
Sunday 9 November 1952	Sasebo Arrived
Tuesday 19 May 1953	Sasebo Departed
Friday 22 May 1953	Hong Kong Arrived
Monday 1 June 1953	Singapore Arrived
Wednesday 17 June 1953	Aden Arrived
Tuesday 23 June 1953	Port Said Arrived
Friday 26 June 1953	Malta Arrived
Friday 3 July 1953	Gibraltar Arrived
Wednesday 8 July 1953	Portsmouth Arrived

Operations in Korean Waters

FIRST TOUR

First Patrol

| Thursday | 26 April | 1951 | left | Sasebo | Japan |
| Monday | 7 May | 1951 | arrived | Sasebo | Japan |

Second Patrol

| Thursday | 10 May | 1951 | left | Sasebo | Japan |
| Sunday | 20 May | 1951 | arrived | Sasebo | Japan |

Third Patrol

| Sunday | 3 June | 1951 | left | Sasebo | Japan |
| Friday | 15 June | 1951 | arrived | Kure | Japan |

Fourth Patrol

| Thursday | 21 June | 1951 | left | Kure | Japan |
| Tuesday | 3 July | 1951 | arrived | Sasebo | Japan |

Fifth Patrol

| Tuesday | 10 July | 1951 | left | Sasebo | Japan |
| Sunday | 22 July | 1951 | arrived | Kure | Japan |

Sixth Patrol

| Tuesday | 24 July | 1951 | left | Kure | Japan |
| Sunday | 5 August | 1951 | arrived | Sasebo | Japan |

Seventh Patrol

Friday	10 August	1951	left	Sasebo	Japan
Saturday	11 August	1951	arrived	Kure	Japan
Sunday	12 August	1951	left	Kure	Japan
Wednesday	21 August	1951	arrived		Okinawa
Thursday	22 August	1951	left		Okinawa
Saturday	25 August	1951	arrived	Kure	Japan

Eighth Patrol

| Friday | 31 August | 1951 | left | Kure | Japan |
| Tuesday | 11 September | 1951 | arrived | Kure | Japan |

Ninth Patrol

Sunday	16 September	1951	left	Kure	Japan
Thursday	27 September	1951	arrived	Kure	Japan
Sunday	30 September	1951	left	Kure	Japan

Operations in Korean Waters

SECOND TOUR

First Patrol

Wednesday	6 February	1952	left	Sasebo	Japan
Friday	15 February	1952	arrived	Sasebo	Japan

Second Patrol

Saturday	23 February	1952	left	Sasebo	Japan
Wednesday	5 March	1952	arrived	Kure	Japan

Third Patrol

Wednesday	12 March	1952	left	Kure	Japan
Sunday	23 March	1952	arrived	Sasebo	Japan

Fourth Patrol

Monday	31 March	1952	left	Sasebo	Japan
Friday	11 April	1952	arrived	Kure	Japan

Fifth Patrol

Thursday	17 April	1952	left	Kure	Japan
Wednesday	30 April	1952	arrived	Sasebo	Japan
Thursday	1 May	1952	left	Sasebo	Japan

Press Release

The British Light Fleet Carrier HMS *Glory* (Captain T.A.K. Maunsell) yesterday returned to Japan on completion of her Second Tour of Operations with the United Nations Forces in Korea.

During the eight months that HMS *Glory* has been operational in this theatre, she has been under the command of Captain K.S. Colquhoun DSO. Captain Maunsell arrived to assume command by air halfway through the ship's last patrol, landing onboard during Flying Operations in a United States Naval Avenger. Captain Colquhoun departed next afternoon in HMNZ Frigate *Rotoiti*.

Since leaving the United Kingdom in January, 1951, HMS *Glory* has steamed some 83,000 miles, operated off four continents, made more than 6,400 aircraft launches and has been screened by more than 60 destroyers from five different nations. The 14th Carrier Air Group (commanded by Leiutenant Commander S.J. Hall, DSC until December 1951, and since then by Lieutenant Commander F.A. Swanton, DSC), consisting of No 804 Sea-Fury Squadron (Lieutenant Commander J.S. Bailey, OBE) and No 812 Firefly Squadron (commanded by the present Air Group Commander until December 1951, and since then by Lieutenant Commander J. Culbertson) has maintained a conspicuously high standard throughout. The group's tasks have been many and varied, sometimes in extreme conditions and often against determined flak opposition, in providing support for ground and naval forces on the west coast of Korea, in harrassing the enemy's lines of communication and attacking his build up areas.

Altogether some 4,839 sorties have been flown on operations, a total of 11,800 flying hours. The cost of this extensive effort has been 27 aircraft lost and more than 140 damaged. The airgroup completed 4,831 deck landings with only 14 accidents, most of the necessary repairs being done onboard. Sixteen pilots have flown more than 130 sorties, Lieutenant K. Whitaker achieving the record number of 149. Whilst 24 aircrew have been recovered from downed aircraft, the ship has suffered the loss of 9 aircrew killed and 1 wounded. The last 939 deck landings were completed without accident.

Among the many tasks that have continually fallen to the lot of the 14th Carrier Air Group are:– close air support to the front line troops, in particular the British Commonwealth Divisions; aircraft spotting for naval bombardment units; photographic missions; armed reconnaissance; air cover and support to naval units defending the coastal islands and strikes against bridges, barracks, store dumps, gun positions and other important targets.

166

In addition HMS *Glory*'s aircraft were responsible for the original sighting of the MIG jet aircraft which crashed off the coast north of Chinnampo last July, and provided air cover for its recovery by HMS *Cardigan Bay*. Also, in September, HMS *Glory* took part in the pounding of Wonsan off the east coast of Korea, with US Cruiser *Toledo* and other bombarding ships. On 17 March this year, HMS *Glory*, wearing the flag of Rear Admiral A.K. Scott Moncrief, reached her peak performance in flying 105 sorties in one day, a total of 203 flying hours.

In all these operations, the ammunition expended on the enemy amounted to over 900,000 rounds of 20mm cannon shells, 14,000 rockets and 3,300 1,000 pound and 500 pound bombs. The damage inflicted on the enemy has been heavy and widespread – the variety and number of targets destroyed and damaged being too numerous to mention in detail. Suffice to say, that as a result of the United Nations' air effort, in which HMS *Glory* has played a considerable part, the enemy have been pre-vented from operating by day and has been forced to lie low until protected by darkness. A considerable effort is made by the enemy to maintain bridges and tunnels in a state of repair, which have been under constant attack. He has gone to intense trouble to camouflage anything of value, and this, together with the increased anti-aircraft defence, has required alertness of a big order on the part of pilots in seeking out suitable targets.

To maintain such a high and sustained effort, serviceability of aircraft is vital. The servicing and maintenance units in HMS *Glory* (Lieutenant Commander (E) I.F. Pearson) have produced a standard that must be hard to equal and, in spite of additional work in repairing battle damage including what is normally considered as only base damage repair work, availability of aircraft has been almost 100%.

Teamwork is a prerequisite to Carrier operations. HMS *Glory*'s motto – Per Concordiam Gloria (Glory Through Unity), has been taken to heart by her ship's company and they can look back with pride at the wonderful performance of her air group, with the knowledge that they too played a vital part, well and truly done.

A tribute must be paid to the three American helicopter units which served HMS *Glory* so faithfully during operations, and who were res-ponsible for so many rescues, often under the most trying conditions. A British helicopter unit has now taken over.

HMS *Glory* leaves tomorrow (Thursday 1 May 1952) for the Mediter-ranean, on relief by HMS *Ocean* (Captain C.L.G. Evans, DSO, DSC).

Ends ...

167

Operations in Korean Waters

THIRD TOUR

First Patrol

Monday	10 November	1952	left	Sasebo	Japan
Thursday	20 November	1952	arrived	Sasebo	Japan

Second Patrol

Friday	28 November	1952	left	Sasebo	Japan
Tuesday	9 December	1952	arrived	Kure	Japan

Third Patrol

Monday	15 December	1952	left	Kure	Japan
Sunday	28 December	1952	arrived	Sasebo	Japan

Fourth Patrol

Sunday	4 January	1953	left	Sasebo	Japan
Tuesday	13 Janaury	1953	arrived	Kure	Japan

Fifth Patrol

Monday	19 January	1953	left	Kure	Japan
Thursday	20 January	1953	arrived	Sasebo	Japan

Sixth Patrol

Thursday	5 February	1953	left	Sasebo	Japan
Tuesday	17 February	1953	arrived	Kure	Japan

Seventh Patrol

Wednesday	25 February	1953	left	Kure	Japan
Sunday	8 March	1953	arrived	Sasebo	Japan

Eighth Patrol

Sunday	15 March	1953	left	Sasebo	Japan
Thursday	26 March	1953	arrived	Kure	Japan

Ninth Patrol

Friday	3 April	1953	left	Kure	Japan
Sunday	12 April	1953	arrived	Sasebo	Japan

Tenth Patrol

Sunday	19 April	1953	left	Sasebo	Japan
Wednesday	29 April	1953	arrived	Kure	Japan

Eleventh Patrol

Monday	4 May	1953	left	Kure	Japan
Sunday	17 May	1953	arrived	Sasebo	Japan
Tuesday	19 May	1953	left	Sasebo	Japan

AIR OFFICE   *Sunday 9 September 1951*

FLYING PROGRAMME

Duty Officer 804 Sqdn –
Duty Officer 812 Sqdn – S/Lieutenant Bates

Event	*Off*	*On*	*A/C*	*Crews*	*Mission*
A.	0615	0815	2 Fu	15-3-4	CAP
			6 Fu	12-3 + DLCOg 99-3	AR or CAS
			1 Fi	22-2	ASP
			4 Fi	23-1-2.22-3-4	AR
B.	0815	1015	2 Fu	15-1-2	CAP
			6 Fu	14 Flt + 99-1-2	AR or CAS
			1 Fi	22-1	ASP
			4 Fi	21 Flt	AR
C.	1015	1215	2 Fu	12-1 + DLCO	CAP
			6 Fu	16 Flt + 12-2-4	AR or CAS
			1 Fi	22-3 + Sherlock	ASP
			4 Fi	23 Flt	AR
D.	1215	1415	2 Fu	14-3-4	CAP
			6 Fu	15 Flt 99-3-4	AR or CAS
			1 Fi	22-4	ASP
			4 Fi	21 Flt	AR
E.	1415	1615	2 Fu	99-1-2	CAP
			6 Fu	12 Flt + Young 14-1-2	AR or CAS
			1 Fi	23-1	ASP
			4 Fi	22-1-2.23-3-4	AR
F.	1615	1815	2 Fu	12-3 + DLCO	CAP
			1 Fi	23-2 + German	ASP
			4 Fi	21-1-2.22-3-4	AR

0515   Flying Stations
0615   Catapult required
       Briefings as usual
       Armament Standard
SCORES: Sat.   51 for Nil
        Sept.   315 for 2
        To date   3,656 for 23
Landings since last accident 80

STAND-BY CAP

Event	*Crews*
A.	12-3 + DLCO
B.	99-3-4
C.	17-1-2
D.	16-3-4
E.	15-3-4
F.	15-1-2

(Signed) S. Keane
Commander (Air)

PATROL	11th
DAY	9th

FLYING PROGRAMME FOR THURSDAY 14 MAY

Duty Officers 801 Mr Lines 821 Lieutenant Banner

Event	Off	On	A/C	Mission	C/S	Crews
A.	0645	0815	2 Fu	CAP	52	Pear Pson
			4 Fu	Strike	55	Hand Bawd Fost Craw
			6 Fi	Strike	76	CO/Agne Spel Garv Hamn Robb Smit
B.	0815	0945	2 Fu	CAP	51	CO Bell
			4 Fu	Strike	53	Leah Ansn Fido Mich
			4 Fu	Strike	54	Macc Hyes Blue Line
			4 Fi	Strike	78	Mars/Cole Bacn Mill Sher
C.	0945	1115	2 Fu	CAP	99	Air Layo
			4 Fu	Strike	52	Pear Pson Hand Bays
			4 Fi	Strike	77	Samp/Hari Spel Robb Hamn Sher
D.	1115	1245	2 Fu	CAP	55.3	Fost Line
			4 Fu	Tarcap	53	Leah Bawd Fido Mich
			4 Fu	Strike	54	Macc Hyes Blue Craw
			4 Fi	Strike	78	Mars/Agne Smit Gari Sher
	1115		1 Fi	Courier		Baco to K16 with Capt Bonfield
E.	1245	1415	2 Fu	CAP	88	Bake Bays
			4 Fu	Strike	51	CO Bell Ansn Pson
			4 Fi	Strike	76	CO/Hunt Mill Samp/West Skin
		1415	Fi	Courier		Baco from K16 with Cmdr

Flying Stations: 0545   *Cat. Required:* 0645   *Briefing:* 30 mins before launch

Fuel: Standard   *Armament:* Fireflies and Furies 2 × 500lb bombs

0730: Gun Functioning   0840: Gun aiming

NOTE: As a precaution against an emergency strike being required after the last land on. 4 Furies armed with bombs will remain at ten minutes notice for launching from 14.30 until 1800. These aircraft should be 4 of those destined for Iwakuni. If they are launched they will land at K16 on completion of the mission, for staging to Iwakuni the following day.

WHEELS UP			
Peangyong Do	0700 – 1000	Moon Rise	New
	1905 – Dusk	Nautical Twilight	0534
Chodo	0700 – 1300	Sunrise	0635
	1905 – Dusk	Sunset	2044
		Nautical Twilight	2143

(Signed) J.W. Sleigh
Commander (Air)

FLYING PROGRAMME – RECORD BREAKER
FOR SUNDAY 5 APRIL

Duty Officers 801 Bawden 821 Lieutenant Agnew

Event	Off	On	A/C	Mission	C/S	Crews
A.	0645	0805	2 Fu	CAP	54	Blue Greg
			4 Fu	Strike	52	Pear Pson Kete Byes
			3 Fu	Strike	53	Leah Fido Mich
			6 Fi	Strike	77	Samp Bann Skin Robb Kent Sher
B.	0805	0925	2 Fu	CAP	88	Bake Bawd
			4 Fu	Strike	51	CO Bell Smit Ansn
			3 Fu	Strike	55	Hand Fost Craw
			6 Fi	Strike	76	CO/West Mill Garv Hamn Mars Grai
C.	0925	1045	2 Fu	CAP	99	Air Kete
			4 Fu	Strike	52	Pear Pson Blue Hyes
			4 Fu	Strike	53	Leah Greg Fido Mich
			6 Fi	Strike	77	Smap Berr Skin Robb Kent Sher
D.	1045	1205	2 Fu	CAP	88	Layo Bell
			3 Fu	AR	51	CO Smit Ansn
			4 Fu	Strike	55	Hand Bawd Fost Craw
			6 Fi	Strike	76	CO/Hari Mill Garv Hamn Mars Grai
E.	1205	1325	2 Fu	CAP	99	Air Pson
			4 Fu	Strike	52	Pear Kete Blue Hyes
			4 Fu	Strike	53	Leah Greg Fido Mich
			4 Fi	Strike	77	Samp Bann Skin Robb
			2 Fi	Tarcap	80	Kent Sher
F.	1325	1445	2 Fu	CAP	88	Bake Fost
			4 Fu	Tarcap	51	CO Bell Smit Ansn
			3 Fu	AR	55	Hand Bawd Craw
			2 Fi	Tarcap	76	CO Mill
			4 Fi	Strike	79	Garv/Cole Hamn Mars Grai
G.	1445	1605	2 Fu	CAP	52	Pear Mich
			3 Fu	Strike	53	Leah Pson Fido
			4 Fu	Strike	54	Blue Hyes Kete Greg
			4 Fi	Strike	77	Samp Bann Kent Berr
			2 Fi	Tarcap	80	Skin Robb
H.	1605	1705	2 Fu	CAP	88	Layo Ansn
			4 Fi	Tarcap	76	CO Mill Garv Hamn
J.	1705	1815	2 Fu	CAP	53	Leah Hyes
			4 Fi	Tarcap	78	Mars/Agne Bann Grai Sher
K.	1845	1925	2 Fu	CAP	51	CO Smit
			2 Fu	Tarcap	55	Hand Fost

Flying Stations: 0545   *Cat. Required:* 0645   *Briefing:* 20 mins before launch

ORDNANCE: Fury 2 × 500lb except on AR and CAP
            Firefly 8 × 60lb R/P throughout
            Bombs Required 104 × 500lb. R/P Required: 384 × 60lb.

FUEL: Fury full internal plus full drop-tanks
      Firefly full internal plus half nacelles

Moon Rise	0107
Nautical Twilight	0625
Sunrise	0723
Sunset	2009
Nautical Twilight	2106

NOTES: 1. No CAS – Gun co-ordination or Courier
            2. Helicopter required after event ABLE

WHEELS UP: Peangyong Do   0945 – 1245
               Chodo   0945 – 1545

(Signed) J.W. Sleigh
Commander (Air)

171

FROM   *GLORY*   UNCLASSIFIED
DTG   050900Z
TO   ADMIRALTY FOR CNI   PRIORITY
INFO   FO2FES (A&A)   CINC FES
            COMNAVFE   CINC MED

Following for Press Release

British naval aircraft from Light Fleet Carrier *Glory* have today Sunday completed 123 operational sorties over North Korea.

2. Starting with only dim light of watery moon pilots of 801 Sea Fury and 821 Firefly Squadrons who have been flying throughout Korean winter maintained continuous effort until mid afternoon so that record for carriers on west coast Korea established by HMS *Ocean* last year should be exactly equalled.

3. Full loads of bombs rockets and cannon shells carried to vital military targets in heart of enemy territory whilst reinforced ammunition supply parties and air engineers in carrier worked against time to prepare aircraft for relaunching.

4. Every aircraft in ship launched alternate flights all pilots completing four sorties totalling more than six hours flying.

5. Throughout this Easter day chaplains were transferred by *Glory*'s helicopter between the Carrier and her United States and Canadian Destroyer escorts where full schedule divine services held.

6. *Glory* now in third year of foreign service commission has already steamed 147 thousand miles almost entirely Korean service her aircraft have completed over twelve thousand five hundred flights including eight thousand four hundred in Korean war.

ORIG (N)   DTG   050900Z
                                             051900K
DIST X XI

5 APRIL 1953

OFFICE   *29 April 1952*

FLYING PROGRAMME
THE LAST DAY! (This time)

Duty Officers 804 Sqdn Lieutenant Scott, Mr Potts
Duty Officers 812 Sqdn S/Lieutenant Cox, S/Lieutenant Cotgrave

Event	Off	On	A/C	Mission	Fuzing
A.	0615	0810	2 Fu	Tarcap	30 sec
			2 Fu	CAP	AB
B.	0800	0955	2 Fu	Tarcap	30 sec
			2 Fu	CAP	AB
			6 Fi	Strike	AB/30 sec (2 to K16 returning 1140)
C.	0945	1140	2 Fu	Tarcap	30 sec
			2 Fu	CAP	AB
			4 Fu	AR	.025
D.	1130	1325	2 Fu	Tarcap	30 sec
			2 Fu	CAP	AB
			6 Fi	Strike	AB/30 sec
E.	1315	1510	2 Fu	Tarcap	30 sec
			2 Fu	CAP	AB
			4 Fu	AR	30 sec
F.	1500	1650	2 Fu	CAP	AB
			2 Fi	Tarcap	AB
			4 Fi	CAS	AB/30 sec
			4 Fu	Strike	.025
G.	1645	1845	2 Fu	CAP	AB
			2 Fu	Tarcap	30 sec

0515 Flying Stations   0615 Catapult required   Briefings as usual

NOTES: 1. One Firefly with 'G' dropper stand-by.
  2. Tarcap will not relieve on station unless ordered.
  3. All flights are to be in the WAITING POSITION FIVE (5) minutes before landing times.
  4. All aircraft except CAP R10 with Bromide Baker 'T'.
  5. All CAP aircraft carry bombs.

EXERCISES: *BROMIDE BAKER TIDES*
Gun functioning   0645 – 0715   High   0921 & 2119
Gun aiming   1045 – 1115   Low   1520
AA Co-ord   1630
SCORES to and including Sunday 27 April:
Sunday   42 for nil
April   726 for 1
To date   6,373 for 36

Landings since last accident = 529

(Signed) S. Keane
Commander (Air)

If you are captured you are required to give the
enemy the information shown on the fol owing
certificate in order that your capture may be
reported to your next of kin. When you are
questioned, but not before, tear off the duplicate
certificate and hand it to the interrogator. DO NOT
ANSWER ANY OTHER QUESTIONS. ALWAYS
CARRY THIS CARD IN ACTION. The interrogator may not take away your S. 1511, S. 43A,
A.B. 64 or A.B. 439. It is important in your own
interest that the particulars of rank should be kept
up to date.

Write in BLOCK CAPITALS

BRITISH FORCES IDENTITY CERTIFICATE

SERVICE
NUMBER...CSKX 891998...RANK...STOKER.........

SURNAME............HYAM.........................

CHRISTIAN OR
FORE NAME(S)......DAVID HAROLD........

DATE OF BIRTH ......12 AUGUST 1932......

- -

BRITISH FORCES IDENTITY CERTIFICATE
(DUPLICATE)

SERVICE
NUMBER...CSKX 891998...RANK.,...STOKER.........

SURNAME............HYAM.........................

CHRISTIAN OR
FORE NAME(S)......DAVID HAROLD........

DATE OF BIRTH ......12 AUGUST 1932......

W.O.P. 33764

FROM   FO 2 FES   UNCLASSIFIED
DTG   160543Z MAY '53
TO   ADMIRALTY FOR CNI   ROUTINE
INFO   C IN C FES *GLORY* FO (AIR) MED: ACNB
NZNB   C IN C MED CANAVHED COM HONG KONG

Following for Press Release Received from *Glory* Begins

HMS *Glory* has just completed longest period naval air operations by any British Commonwealth Carrier in Korean campaign.

2. Since leaving UK January 1951 ship has spent 530 days at sea and steamed 157,000 miles. During this period *Glory* has completed 15 months war service and spent 316 days at sea in Korean waters. Of a total 13,700 flights from carriers deck more than 9,500 have been operational sorties over northern Korea. A constant toll of enemy troops, disruption and destruction of rail and road communications and wide range of military targets have continued. During two Korean winters ice and snow has been repeatedly had to be cleared from flight deck and aircraft but operations have been sustained in some of worst weather ever experienced by naval air crew.

3. *Glory*'s Firefly and Seafury aircraft have been surprising rail and road transport by night in northern Korea whilst a notable feature of daylight operations has been close air support for the British Commonwealth division at the front line.

4. Extent of United Nations organisation borne by fact that nearly 100 destroyers of Royal Navy, United States, Royal Australian, Canadian and Netherlands navies have operated as escorts to the British Carrier.

5. On her last day in Korean waters before returning to England ship had a Memorial Service. During past six months 12 members of *Glory*'s aircrew have flown on missions from which they have not returned. Ends.

DIST   X 3 5 C Cl D E H G I L M N O S Sl 3 T V A 2 7 8 FL LOG
                                             DTG   160543Z

FE 676   P/L   TOT   1651   CFR   17.5.53

Korean Operations

Monday 23 April 1951 to Sunday 30 September 1951

14th CARRIER AIR GROUP
Lieut-Cmdr (P) S.J. Hall DSC RN (Air Group Commander)
Lieut-Cmdr (P) I.F. Pearson RN (Air Engineering Officer)
Lieut (E) R.M. Fillery RN

804 NAVAL AIR SQUADRON
Lieut-Cmdr (P) J.S. Bailey OBE RN (Commanding Officer)
Lieut-Cmdr (P) M.A. Birrell RN (Senior Pilot)

Lieutenant (P)	R.H.	Kilburn
	I.W.	Campell
	G.W.	Bricker
	K.	Whitaker
	J.A.	Winterbotham
	W.R.	Hart
	R.C.B.	Trelawney
	D.A.	McNaughton
	P.G.	Young
	R.F.	Hubbard
	P.S.	Davis
	J.R.	Fraser
	E.P.L.	Stephenson
	A.	Fane
	P.A.L.	Watson
	P.	Barlow
Sub-Lieutenant (P)	J.R.	Howard
Commissioned Pilot	F.	Hefford
	P.	MacKerral
	W.A.	Newton
	R.E.	Collingwood
	P.J.	Potts
	T.	Sparke
	C.E.	Mason
	P.O.	Richards
	M.I.	Darlington
	D.F.	Fieldhouse

Korean Operations

Monday 23 April 1951 to Sunday 30 September 1951

14th CARRIER AIR GROUP

812 NAVAL AIR SQUADRON
Lieut-Cmdr (P) F.A. Swanton DSC RN (Commanding Officer)
Lieut-Cmdr (P) R.H.W. Blake RN (Senior Pilot)

Lieutenant (P)	T.V.G.	Binney
	R.A.L.	Smith
	W.H.	Gunner
	P.G.W.	Morris
	T.G.	Davies
	R.E.	Wilson
	J.K.	Arbuthnot
	P.A.	Jordan
	J.H.	Sharp
	R.	Williams
	D.E.	Johnson
Lieutenant (O)	J.G.C.	Harvey
	G.E.	Legg
	W.J.	Carter
	A.D.	Hooper
Sub-Lieutenant (P)	J.S.	Tait
	L.R.	Shepley
Sub-Lieutenant (O)	R.J.	Bates
Commissioned Pilot	J.A.	Neilson
	J.P.	Hack
	J.T.	Griffiths
	M.H.C.	Purnell
	R.G.	Clarke
	W.F.	Cockburn
	J.H.	Eagle
	C.B.	Sleight
Pilot 3	S.W.E.	Ford
Aircrewman	K.L.J.	Sims
	G.B.	Wells
	G.	Mortimer
	D.E.	Jackson

SEARCH AND RESCUE HELICOPTER

Lieutenant	P.	O'Mara, United States Navy
CPO		Fridley, United States Navy

177

Korean Operations
Sunday 27 January 1952 to Monday 5 May 1952

14th CARRIER AIR GROUP
Lieut-Cmdr (P) F.A. Swanton DSC RN (AGC)
Lieut-Cmdr (P) I.F. Pearson RN (AEO)
Lieut R.M. Fillery RN (Air Engineer)
Lieut C.M. Caldecott RN (Air Engineer)

804 NAVAL AIR SQUADRON
Lieut-Cmdr (P) J.S. Bailey OBE RN (CO)
Lieut-Cmdr (P) M.A. Birrell RN (Senior Pilot)

Lieutenant (P)	J.R.	Fraser
	K.	Whitaker
	D.A.	McNaughton
	P.G.	Young
	P.S.	Davis
	N.E.	Peniston-Bird
	R.J.	Overton
	M.E.	Scott (Royal Australian Navy)
	A.G.	Cordell (Royal Australian Navy)
	P.A.L.	Watson
	P.	Barlow
	P.I.	Normand
Sub-Lieutenant (P)	J.R.	Howard
	D.L.G.	Swanson
	C.E.	Haines
	P.H.	Wyatt (Royal Australian Navy)
	A.G.	Powell (Royal Australian Navy)
Commissioned Pilot	M.I.	Darlington
	D.F.	Fieldhouse
	F.	Hefford
	W.A.	Newton
	R.E.	Collingwood
	B.J.	Potts
	A.F.	Griggs

SEARCH AND RESCUE HELICOPTER

Lieutenant (P)	E.S.	Taylor RN
	C.W.	Perry RN

Korean Operations

Sunday 27 January 1952 to Monday 5 May 1952

14th CARRIER AIR GROUP

812 NAVAL AIR SQUADRON
Lieut-Cmdr (P) J.M. Culbertson RN (CO)
Lieut (F) J.R. Hone RN (Senior Pilot)

Lieutenant (P)	T.J.	Kinna
	P.B.	Reynolds
	W.LeG.	Jacob
	E.J.	Meadowcroft
	J.G.	Pope
Lieutenant (O)	R.C.	Hubbard
	A.D.	Hooper
	C.J.	Fursey
Sub-Lieutentant (P)	J.M.	Wood
	J.S.	Tait
	J.S.	Cotgrove
	R.	Cox
Sub-Lieutenant (O)	R.J.	Bates
	M.C.S.	Apps
	J.S.	Kendall
	M.J.	Jenvey
Commissioned Pilot	R.G.	Clarke
	J.T.	Griffiths
	M.H.C.	Purnell
	C.B.	Sleight
Commissioned Observer	G.G.	Gibbs
Aircrewman	T.	Leigh
	A.	Japp
	L.M.	Edwards
	L.J.	Stevens

Korean Operations

Saturday 8 November 1952 to Tuesday 19 May 1953

801 NAVAL AIR SQUADRON

Commander (P) B.C.G. Place VC DSC RN
Lieut-Cmdr (P) P.B. Stuart (Commanding Officer)

Lieutenant (P)	J.H.S.	Pearce (Senior Pilot)
	P.D.	Handscombe
	R.J.	McCandless
	A.J.	Leahy
	J.H.	Fiddian-Green
	A.R.	Graham
	R.	Nevill-Jones
	P.	Wheatley
	J.R.T.	Bluett
	P.A.B.	Wemyss
	E.R.	Anson
	V.B.	Mitchell
	J.A.S.	Crawford
	C.A.	McPherson
	J.	Bawden
(E)(A/E)(P)	D.G.	Mather
Sub-Lieutenant (P)	B.E.	Rayner
	R.D.	Bradley
	M.B.	Smith
	D.McL.	Baynes
	W.J.B.	Keates
	G.B.S.	Foster
	J.F.	Belville
	M.	Hayes
	A.R.	Pearson
Lieutenant (A)	T.O.	Adkin RNVR
	J.C.R.	Buxton RNVR
Sub-Lieutenant (A)	J.M.	Simmonds RNVR
Commissioned Pilot	P.R.	Lines
Lieutenant (E)(A/E)	R.A.	Langley (Air Engineering Officer)
SCLO (Air)	F.E.	Lemming (ALO)

180

Korean Operations

Saturday 11 November 1952 to Tuesday 19 May 1953

821 NAVAL AIR SQUADRON
Lieut-Cmdr (P) J.R.N. Gardner (Commanding Officer)
Lieut (P) P. Cane (Senior Pilot)

	P.	Dallosso
	G.D.H.	Sample
	R.E.	Barrett
	R.	Garvin
	W.R.	Sherlock
	A.L.L.	Skinner
	J.M.	Bacon
	J.F.	McGrail
	J.G.	Marshall
	W.R.	Heaton
	P.	Spelling
	H.J.	Smith
	B.V.	Bacon
	A.G.	Hamon
	P.G.	Fogden
	P.R.	Banner
	P.	Millett
(E)(A/E)(P)	D.F.	Robbins
(E)(A/E)(O)	J.M.	Hunter (Senior Observer)
	J.S.	Agnew
Sub-Lieutenant (P)	J.R.DeB.	Wailes
Sub-Lieutenant (O)	J.R.	Coleman
	R.	Harrison
	D.J.R.	West
Commissioned Pilot	M.	Kent
Lieutenant (E)(A/E)	J.R.P.	Lansdown (Air Engineering Officer)

AIR AND RESCUE HELICOPTER

Lieutenant (P)	A.P.	Daniels DSM RN
	G.	Spetch RN
Aircrewman	E.R.	Ripley
	J.	Glenn

Korean Operations

HONOURS AND AWARDS

Capt Kenneth Stuart Colquhoun DSO	CBE
Capt Edgar Duncan Goodenough Lewin DSO DSC★	CBE
Cmdr (P) James Wallace Sleigh DSO DSC	OBE
Cmdr Robert Love Alexander DSO DSC	Mention in Dispatches
Cmdr (S) William Hugh Field	Mention in Dispatches
Lieut-Cmdr (P) Sidney James Hall DSC	DSO
Lieut-Cmdr (P) Francis Alan Swanton DSC★	DSO
Lieut-Cmdr (P) John Savile Bailey OBE	DSC
Lieut-Cmdr (P) James Robert Nigel Gardner	DSC
Lieut-Cmdr (P) Reginald Howard Watson	DSC
Lieut-Cmdr (P) Peter Basil Stuart	DSC
Lieut-Cmdr (P) William Thomas Rutherford Smith	MBE
Lieut-Cmdr (P) Ian Francis Pearson (E) (A/E)	MBE
	Mention in Dispatches
Lieut-Cmdr (O) Philip Reginald Spademan	Mention in Dispatches
Lieut-Cmdr (O) John Gabriel Cavendish Harvey	DSC
Lieut (P) Alan John Leahy	DSC
Lieut (P) Douglas Arthur McNaughton	DSC
Lieut (P) Robert John McCandless	DSC
Lieut (P) Paul Millett	DSC
Lieut (P) James Henry Silvester Pearce	DSC
Lieut (P) Geoffrey David Hutton Sample	DSC
Lieut (P) Victor Giles Binney	Mention in Dispatches
Lieut (P) Robin Christopher Beaumont Trewlawney	Mention in Dispatches
Lieut (P) Roi Egerton Wilson	Mention in Dispatches
Lieut (O) Anthony Desmond Hooper	Mention in Dispatches
Lieut (E) Peter Barlow (P) (A/E)	DSC
Lieut (E) Derek Graham Mather	Queen's Commendation (PoW)
Lieut Paul O'Mara United States Navy	Hon MBE
Cmmd Pilot Frederick Hefford	DSC
	Mention in Dispatches

Korean Operations

HONOURS AND AWARDS

Cmmd Pilot John Alexander Neilson	DSC
Cmmd Pilot Maurice Henry Charles Purnell	DSC
Cmmd Pilot Michael Ian Darlington	Mention in Dispatches
Cmmd Pilot Derek Frederick Fieldhouse	Mention in Dispatches
Cmmd Air (E) Sydney Jones BEM	Mention in Dispatches
Cmmd Air (E) James Henry Freeland (O)	Mention in Dispatches
Bandmaster Walter James Spencer Royal Marines	Mention in Dispatches
Chief Air Art Arthur Charles Fooks	BEM
Chief ERA George Jack Turp	BEM
Chief Air (E) Ronald Cater	BEM
AA2 Leslie Green	BEM
AA3 John Stanley Abbott	BEM
AA3 Gerald Peter Chisholm	BEM
Chief Yeoman of Signals Kenneth Charles Youngjohns	Mention in Dispatches
Chief Petty Officer (Tel) Thomas Edward Carlow	Mention in Dispatches
Chief Petty Officer (Air) Albert Sadler	Mention in Dispatches
Chief ERA Michael Conheeney	Mention in Dispatches
Chief Air Artificer (O) Gerald Wright Jones	Mention in Dispatches
Chief ERA Alfred James Brett	Mention in Dispatches
AA3 Leonard Matthew Mitchell	Mention in Dispatches
AA3 Harold Sidney Tuffin	Mention in Dispatches
Chief Electrician (A) Ronald Albert Edward Morris	Mention in Dispatches
Petty Officer Charles McKiddie	Mention in Dispatches
Petty Officer (A) Horace Dowler	Mention in Dispatches
Petty Officer (A) James Saunders Leitch	Mention in Dispatches
Petty Officer Stoker Mechanic John Smith	Mention in Dispatches
Leading Airman Alexander Jones Law	Mention in Dispatches
Leading Airman (PM) Kenneth Arthur McMichael	Mention in Dispatches
Leading Airman (A) Albert Mark	Mention in Dispatches
Leading Airman (E) Peter Ernest Jones	BEM
Leading Airman (O) Ronald Daily	Mention in Dispatches
Electrician (Air) Herbert Geoffrey Brice	BEM

Roll of Honour

HMS *GLORY* KOREA 1951–1953

Lieutenant Edward Peter Langdale Stephenson	28 April	1951
Pilot (3) Stanley William Edwin Ford	5 June	1951
Lieutenant (P) John Harry Sharp	28 June	1951
Aircrewman (I) George Bertram Wells	28 June	1951
Lieutenant (P) Robert Williams	16 July	1951
Sub-Lieutenant (O) Ian Robertson Shepley	16 July	1951
Commissioned Pilot Terence William Sparke	18 July	1951
Sub-Lieutenant (O) Ronald George Albert Davey	22 Sept	1951
Lieutenant (P) Richard James Overton	15 March	1952
Lieutenant (P) Richard Nevill-Jones	18 Nov	1952
Petty Officer (Air) Victor Colman	19 Nov	1952
Lieutenant (P) Alan Philip Daniels	16 Dec	1952
Aircrewman (I) Ernest Raymond Ripley	16 Dec	1952
Lieutenant (P) Peter George Fogden	20 Dec	1952
Lieutenant (P) Robert Edward Barrett	25 Dec	1952
Sub-Lieutenant (P) Brian Edward Rayner	5 Jan	1953
Sub-Lieutenant (P) Janes Malcolm Simonds	5 Jan	1953
Lieutenant (P) Cedric Alexander MacPherson	11 Feb	1953
Sub-Lieutenant (P) Richard Derek Bradley	14 Feb	1953
Lieutenant (P) John Thomas McGregor	25 April	1953
Sub-Lieutenant (P) Walter John Butler Keates	25 April	1953

Also during the commission:

ERA3 D.E. Dixon C/MX102140	2 Oct	1951
NA1 T.J. Williams L/SFX847695	27 Dec	1951
EA4 E. Senior C/MX667728	21 Sept	1952

Through the Hell of Wonsan

West Coast easy we'd been told,
On the East you'll not be bold.

Rings a twitching, jaws hung slack
So we went in to attack.
Mouthing prayers to all and sundry
Waiting for the awful flak.

Nothing happened, all was quiet,
Of flak there came up not a puff.
We created quite a riot
Cannons, rockets, bombs and stuff.

Firefly leader feeling bolder
Chose a target that was older.
Seeing nothing much was doing
Blasted pin-point ancient ruin.
(Relic of Korean culture vandalised by modern vulture.)

The worst thing for us hey-ho
We had to use that bloody RATO!

A Day on the Flight Deck

Before the 'Wakey, Wakey' goes
The Air Department's on its toes
Preparing for the great affray
That's just another working day
For *Glory*'s mighty aircraft.

> They're naval airmen and pilot's mates
> Responsible for fuel states
> And other tasks that go to make
> The pilot's job a 'piece of cake'
> With *Glory*'s mighty aircraft.

They're always working with a smile
And have to be quite versatile.
'Munitioning ship they have to do
As well as being maintenance crew.
For *Glory*'s mighty aircraft.

> The pilots come, some short some tall,
> In answer to the tannoy's call.
> Some unknown, some of repute
> Climb up and fix their parachutes
> In *Glory*'s mighty aircraft.

They've started up; before they go
They've got to check the radio
On channel Able and channel Baker,
If they don't work then blame the maker
Of *Glory*'s mighty aircraft.

> The pilots sit there cold and glum
> With engines warm and revs quite steady
> Until at last the famous thumb
> Says bombs and rockets now are ready
> On *Glory*'s mighty aircraft.

The chocks away, they taxi down,
All hoping not to cause a frown
On Zeus's brow; the Thunderer
And undisputed Arbiter
Of *Glory*'s mighty aircraft.

With ERA's and stokers too
The catapult is kept like new
It's working hard for every range
Though slight delays are never strange
To *Glory*'s mighty aircraft.

And now with every aircraft airborne
They quickly on their leader form
To find and strike with all their might
The target and all else in sight
They're *Glory*'s mighty aircraft.

The time for landing on is ripe
They're standing by to make the pipe:
Smoking restrictions throughout the Carrier
An aircrafts just picked up the Barrier.
It's *Glory*'s mighty aircraft.

And now this short patrol is o'er
We turn around and head for shore
We'll have to stop on route to Kure
For we must change at least one Fury
For *Glory*'s mighty aircraft.

D.J.C. C.P.O. 67 Mess

A Royal and His Gun

As into the stormy Yellow Sea,
Slowly sinks the setting sun,
Aboard the Carrier *Glory*
A bootneck cleans his gun.

'For months I've cleaned this ruddy gun'
Was his sad and doleful cry,
'And as for targets hostile
They have always passed us by.

'The Furies and the Fireflies
Daily, bravely have a go,
Releasing bombs and rockets
On the "Chicoms" down below.

'And as all day I sit and dream
though often I only sits;
Over yonder is Korea
They are blowing blokes to bits.

'As once again they come in sight,
In formation, close and neat,
How I envy those who combat,
Sitting in a pilot's seat!

'Bad approach, wave off, a floater,
Over the side he disappears,
From the wreckage crawls the pilot,
Not for him a widow's tears.

'Dear old Bofors, how I love you
Quoth the Royal tenderly,
You don't fly around the heavens
Ending in the deep blue sea.

'Only fools and birds go flying,
Is what I was always taught.
In your shade I'll write my letters
Of these battles being fought.

'And when at last I leave Korea
With my medals, two or three,
I'll remember with much longing
Of those nights I slept with thee.'

JG

188

Up Spirits

In the days of Admiral Nelson,
Or it may have been before,
The Navy got its heritage,
Its customs and its lore.

Now some of these were good things,
And some of them were not
But they'll never find one better
Than that little daily tot.

It isn't served haphazardly
Like tea or even beer,
But with pencil, book, and measure,
And other essential gear.

Far below in secret vault,
Strange vessels gleaming bright,
Jack Dusty and his retinue
Perform the sacred rite.

With murmured incantations,
Like 'Twenty Mess, stop two!'
The High Priest and his acolyte
Dispense the holy brew.

With bottle, jug etcetera,
We muster at the shrine,
'Hands off you thieving B.......,
That flippin' Fanny's mine!'

When the sea get rough and stormy
And you feel you've had enough,
When the chef has burnt the 'taters'
And the OD's dropped the duff.

When your boots are full of 'oggin',
And you're full right up to here,
There's nowt can cheer you half as much
As that daily little tot.   R.D., Leading Signalman

1952

A year that hasn't been so bad,
By all, some good times have been had.
On seventh of Jan we said goodbye
And left the Sydney girls to cry.
The next three months, were you know where,
Breathing deep the Korean air.
Came merry May and we were bound
For Malta, instead of Plymouth Sound.
In June the ship was high and dry,
And our old Air Group said goodbye.
Then in July we sailed away,
Port Said, and back to Bighi Bay.
But ere the month had past away,

To Istanbul for a short stay.

Strong man Nequib then cast his net,
And for Tobruk our course was set.
The next three weeks at Larnaca,
Then followed a spell at Malta.
'Twas on this trip before the mast,
The hundredth thousand mile went past.

Barcelona in September,
Senoritas, you remember.

Then came the day October nine,
When we sailed east in weather fine.
November found us once again
Amidst Korean snow and rain.
And December fifty-one,
Christmas and New Year's Eve in one.
The Daily Mail has kindly sent
A bar of nutty, fags and beer
To help increase our Christmas cheer.
Thus in the year left in our wake
As on this day we eat our cake.
Eat up, drink deep, enjoy yourself,
The year has gone, 'tis on the shelf

'Farewell' to 1952.

GLORY MORNING STAR 31 December 1952

No 837 Naval Air Squadron

BASES 1945 TO 1947

Ayr (HMS *Wagtail*)	2 April	1945
HMS *Glory*	4 April	1945
Ayr (HMS *Wagtail*)	18 April	1945
HMS *Glory*	11 May	1945
Dekheila (HMS *Grebe*)	24 May	1945
HMS *Glory*	25 June	1945
Katurkuranda (HMS *Ukussa*)	15 July	1945
HMS *Glory*	27 July	1945
Schofields (HMS *Nabthorpe*)	16 Aug	1945
HMS *Glory*	1 Sept	1945
Jervis Bay (HMS *Nabswick*)	11 Sept	1945
Nowra (HMS *Nabbington*)	29 Oct	1945
HMS *Glory*	14 Jan	1946
Williamtown	15 Feb	1946
HMS *Glory*	10 June	1946
Trincomalee	9 Aug	1946
HMS *Glory*	20 Sept	1946
Kai Tak	1 Oct	1946
HMS *Glory*	4 Nov	1946
Sembawang	18 Nov	1946
HMS *Glory*	9 Dec	1946
Kai Tak	19 Dec	1946
HMS *Glory*	14 Feb	1947
Trincomalee	27 Feb	1947
HMS *Glory*	15 April	1947
Trincomalee	30 April	1947
Sembawang	17 May	1947
HMS *Glory*	19 June	1947
Squadron disbanded UK	6 Oct	1947

COMMANDING OFFICERS

Lieut- Cmdr R.B. Martin RNVR	1 Aug	1944
Lieut-Cmdr W. Siddall-Simpson RNVR	14 Dec	1945
Lieut-Cmdr G. Hamilton-Bates RN	25 Jan	1946
Lieut-Cmdr R.H. Hain RN	7 Mar	1947

AIRCRAFT
Barracuda II
Firefly I

1831 Naval Air Squadron

BASES 1945 TO 1947

HMS *Glory*	11 May	1945
Hal Far	22 May	1945
Dekheila (HMS *Grebe*)	18 June	1945
HMS *Glory*	2 July	1945
Katurkuranda (HMS *Ukussa*)	15 July	1945
HMS *Glory*	27 July	1945
Schofields (HMS *Nabthorpe*)	16 Aug	1945
HMS *Glory*	1 Sept	1945
Jaquinot Bay (Detached)	30 Sept	1945
Nowra (HMS *Nabbington*)	15 Nov	1945
HMS *Glory*	19 Jan	1946
Williamtown	15 Feb	1946
HMS *Glory*	10 June	1946
Trincomalee	15 July	1946
HMS *Vengeance*	16 July	1946
Squadron disbanded	13 Aug	1946

COMMANDING OFFICERS

Lieut-Cmdr R.W.M. Walsh RN	1 Nov	1944
Lieut-Cmdr R.T. Leggot MBE RN	28 March	1946

AIRCRAFT
Corsair IVs

192

No 806 Naval Air Squadron

BASES 1946 TO 1947

Trincomalee	13 May	1946
HMS *Glory*	20 Sept	1946
Kai Tak	1 Oct	1946
HMS *Glory*	6 Nov	1946
Sembawang	18 Nov	1946
HMS *Glory*	6 Dec	1946
Kai Tak	19 Dec	1946
HMS *Glory*	14 Feb	1947
Sembawang	17 May	1947
HMS *Glory*	19 June	1947
Squadron disbanded	6 Oct	1947

COMMANDING OFFICERS

Lieut-Cmdr A.W. Bloomer RN	1 Jan	1946
Lieut-Cmdr R.P. Thurston RN	1 Oct	1946
Lieut-Cmdr W.N. Waller RN	1 Jan	1947

AIRCRAFT

Seafire XV

193

No 804 Naval Air Squadron

BASES 1949 TO 1952

Hal Far (HMS *Falcon*)	18 July	1949
HMS *Glory*	20 Dec	1949
Hal Far (HMS *Falcon*)	31 March	1950
Deversoir (Detached)	17–23 March	1950
Nicosia (Detached)	21 April	1950
Hal Far (HMS *Falcon*)	5 May	1950
HMS *Glory*	16 May	1950
Hal Far (HMS *Falcon*)	31 July	1950
HMS *Glory*	4 Sept	1950
Hal Far (HMS *Falcon*)	1 Oct	1950
Castel Benito (Detached 8 days)	7 Nov	1950
HMS *Glory*	8 March	1951
Sembawang	8 Oct	1951
HMS *Glory*	11 Oct	1951
Nowra (HMS *Albatross*)	22 Oct	1951
HMS *Glory*	2 Jan	1952
HMS *Theseus*	26 May	1952

COMMANDING OFFICERS

Lieut-Cmdr (A) C.F. Hargreaves RN	6 Feb	1949
Lieut-Cmdr J.S. Bailey OBE RN	1 Dec	1950

AIRCRAFT
Sea Fury FB11

No 812 Naval Air Squadron

BASES 1949 TO 1952

HMS *Glory*	11 Nov	1949
Hal Far (HMS *Falcon*)	28 Nov	1949
HMS *Glory*	19 Dec	1949
Hal Far (HMS *Falcon*)	31 March	1950
HMS *Glory*	12 May	1950
Hal Far (HMS *Falcon*)	6 June	1950
HMS *Glory*	12 June	1950
Hal Far (HMS *Falcon*)	31 July	1950
HMS *Glory*	17 Aug	1950
Hal Far (HMS *Falcon*)	1 Oct	1950
HMS *Glory*	10 Oct	1950
Castel Benito	8 Nov	1950
HMS *Glory*	22 Nov	1950
Hal Far (HMS *Falcon*)	2 Feb	1951
HMS *Glory*	8 March	1951
Nowra (HMS *Albatross*)	12 Nov	1951
HMS *Glory*	10 Jan	1952
HMS *Theseus*	26 May	1952

COMMANDING OFFICERS

Lieut-Cmdr R.M. Fell RN	6 March	1949
Lieut-Cmdr R.G. Hunt RN	17 July	1950
Lieut-Cmdr F.A. Swanton DSC RN	1 March	1951
Lieut-Cmdr J.M. Culbertson RN	18 Dec	1951

AIRCRAFT
Firefly FR5

No 821 Naval Air Squadron

BASES 1952 TO 1953

Hal Far (HMS *Falcon*)	26 June	1952
HMS *Glory*	2 Sept	1952
Kai Tak (HMS *Tamar*) Disbanded	25 May	1953

COMMANDING OFFICER

Lieut-Cmdr J.R.N. Gardner RN	12 May	1952

AIRCRAFT
Firefly FR5

No 801 Naval Air Squadron

BASES 1952 TO 1954

Hal Far (HMS *Falcon*)	24 June	1952
HMS *Glory*	2 Sept	1952
HMS *Ocean*)	18 May	1953
Lossiemouth (HMS *Fulmar*)	7 July	1953
HMS *Illustrious*	31 Aug	1953
Lee-on-Solent (HMS *Daedalus*)	28 Sept	1953
HMS *Glory*	2 Oct	1953
Hal Far (HMS *Falcon*)	12 Nov	1953
HMS *Glory*	18 Nov	1953
Hal Far (HMS *Falcon*)	18 Dec	1953
HMS *Glory*	4 Jan	1954
Lee-on-Solent (HMS *Daedalus*)	28 Feb	1954

COMMANDING OFFICERS

Lieut-Cmdr P.B. Stuart RN	1 May	1952
Lieut-Cmdr J.H.S. Pearce DSC RN	1 March	1954

AIRCRAFT
Sea Fury FB11

No 826 Naval Air Squadron

BASES 1953 TO 1954

Lee-on-Solent (HMS *Daedalus*)	2 Oct	1953
HMS *Glory*	2 Nov	1953
Hal Far (HMS *Falcon*)	12 Nov	1953
HMS *Glory*	18 Nov	1953
Hal Far (HMS *Falcon*)	18 Dec	1953
HMS *Glory*	4 Jan	1954
Lee-on-Solent (HMS *Daedalus*)	28 Feb	1954

COMMANDING OFFICER

Lieut-Cmdr J.W. Powell DSC RN	4 Dec	1952

AIRCRAFT
Firefly AS6

No 807 Naval Air Squadron

BASES 1952

HMS *Glory*	6 July	1952
Hal Far (HMS *Falcon*)	8 July	1952
HMS *Glory*	21 July	1952
Hal Far (HMS *Falcon*)	13 Aug	1952

COMMANDING OFFICER

Lieut-Cmdr A.J. Thomson DSC RN	15 June	1951

AIRCRAFT
Sea Fury IIs

No 810 Naval Air Squadron

BASES 1952

HMS *Glory*	21 May	1952
Kasfareet	23 June	1952
HMS *Glory*	6 July	1952
Hal Far (HMS *Falcon*)	14 Aug	1952

COMMANDING OFFICERS

Lieut-Cmdr D.E. Johnson RN	29 June	1951
Lieut-Cmdr A.W. Bloomer RN	28 June	1952

AIRCRAFT
Firefly FR5

APPENDIX 50

No 898 Naval Air Squadron

BASES 1952

HMS *Glory*	6 July	1952
Hal Far (HMS *Falcon*)	8 July	1952
HMS *Glory*	21 July	1952
Hal Far (HMS *Falcon*)	15 Aug	1952

COMMANDING OFFICER

Lieut-Cmdr T.L.M. Brander DSC RN	4 July	1951

AIRCRAFT
Sea Fury FB11

807, 810 and 898 Squadrons embarked in *Glory* for the visit to Istanbul, Turkey in July 1952.

Winter Cruise
(October 1953 to March 1954)

Monday	26 October	1953	Departed	Rosyth
Wednesday	28 October	1953	Arrived	Portsmouth
Monday	2 November	1953	Departed	Portsmouth
Thursday	5 November	1953	Arrived	Devonport
Friday	6 November	1953	Departed	Devonport
Monday	9 November	1953	Arrived	Gibraltar
Monday	9 November	1953	Departed	Gibraltar
Friday	13 November	1953	Arrived	Malta
Wednesday	20 January	1954	Departed	Malta
Thursday	21 January	1954	Arrived	Naples
Wednesday	27 January	1954	Departed	Naples
Friday	29 January	1954	Arrived	Malta
Monday	15 February	1954	Departed	Malta
Wednesday	17 February	1954	Arrived	Villefranche
Saturday	20 February	1954	Departed	Villefranche
Tuesday	23 February	1954	Arrived	Gibraltar
Thursday	25 February	1954	Departed	Gibraltar
Sunday	28 February	1954	Arrived	Falmouth
Sunday	28 February	1954	Departed	Falmouth
Monday	1 March	1954	Arrived	Portsmouth

826 and 801 Naval Air Squadrons joined HMS *Glory* on Monday 2 November 1953 and disembarked to Royal Naval Air Station Lee-on-Solent Sunday 28 February 1954. Both squadrons spent Christmas 1953 ashore at Royal Naval Air Station, Hal-Far, Malta (HMS *Falcon*).

FERRYING DUTIES TO THE FAR EAST

Left	Portsmouth	Monday	13 September	1954
Arrived	Glasgow	Wednesday	15 September	1954
Left	Glasgow	Friday	17 September	1954
Arrived	Gibraltar	Tuesday	21 September	1954
Left	Gibraltar	Tuesday	21 September	1954
Arrived	Malta	Friday	24 September	1954
Left	Malta	Saturday	25 September	1954
Arrived	Port Said	Tuesday	28 September	1954
Left	Port Said	Tuesday	28 September	1954
Arrived	Aden	Tuesday	5 October	1954
Left	Aden	Tuesday	5 October	1954
Arrived	Trincomalee	Tuesday	12 October	1954
Left	Trincomalee	Friday	15 October	1954
Arrived	Singapore	Wednesday	20 October	1954
Left	Singapore	Saturday	30 October	1954
Arrived	Colombo	Thursday	4 November	1954
Left	Colombo	Saturday	6 November	1954
Arrived	Aden	Friday	12 November	1954
Left	Aden	Saturday	13 November	1954
Arrived	Malta	Monday	22 November	1954
Left	Malta	Friday	26 November	1954
Arrived	Gibraltar	Monday	29 November	1954
Left	Gibraltar	Thursday	2 December	1954
Arrived	Portsmouth	Monday	6 December	1954

Captain H.W. Sims-Williams, Royal Navy

THE FINAL VOYAGE

Tuesday 8 May 1956	Rosyth Departed
Sunday 13 May 1956	Portland Arrived
Monday 14 May 1956	Devonport Arrived
Monday 11 June 1956	Milford Haven Arrived
Friday 15 June 1956	Rosyth Arrived
Friday 22 June 1956	Paid Off

Captain T.N. Masterman OBE, Royal Navy

HMS *GLORY*'S AIRCRAFT

Fairey Barracuda: The first aircraft to land aboard *Glory* were the Torpedo Bomber/Reconnaissance Barracuda IIs of 837 Naval Air Squadron. They carried a crew of three: Pilot, Observer and Telegraphist Air Gunner. Barracudas could be armed with torpedos, depth charges or bombs. They had a wingspan of 49ft 2in, length of 39ft 9in and had a maximum weight of 14,250lbs. Powered initially by Rolls Royce Merlin engines, later types were fitted with Griffon engines giving improved performance and a maximum 315 knots. Despite a disappointing performance as a dive-bomber and having a rather odd looking appearance with its high wing and big windows underneath, the Barracuda was nevertheless used extensively in all theatres during WW2 and is best remembered for its attack on the German battleship *Tirpitz*.

Chance Vought Corsair: Nearly 2,000 Corsairs were received by the Fleet Air Arm during WW2. 1,831 Naval Air Squadron joined *Glory* with 21 Corsair IVs in May 1945. Single seat aircraft, with a top speed of 365 knots, they could perform the role of Fighter or Fighter Bomber. Armed with four 0.5 in Browning machine guns in the wings they could also carry two 1,000lb or 500lb bombs. 100 gallon drop tanks could also be fitted. The Corsair was a stylish aeroplane and powered by its 2,250hp Pratt and Whitney radial engine carried out its various tasks with distinction. Many of those on loan to the Royal Navy were ditched at sea after the war. Corsairs were operational with the US and French navies until the 1950s.

Supermarine Seafire: There were many Seafire variants and when 806 Naval Air Squadron joined *Glory* in 1946 it was equipped with Seafire FXVs powered by Rolls Royce Griffon engines. It was undoubtedly the best British built single seat fighter at that time. Armed with four 20mm cannon, it could be adapted to carry rockets, bombs and underwing drop tanks. Maximum speed was about 400 knots. It was, however, notorious for its disconcerting behaviour when landing on a carrier, problems caused by a narrow and not very robust undercarriage, and by the pilot's lack of forward vision on final approach. Despite the deck landing problem Seafires gave many years of valuable service to the Fleet Air Arm.

Fairey Firefly: *Glory*'s 837 Squadron were re-equipped with Firefly F1s whilst in Australia in October 1945. Other Firefly squadrons to serve in *Glory* were 812, 821, 826 and 810. The Firefly was the last in a long line of piston engined, two seater, reconnaissance fighters with the Fleet Air Arm. It could be argued that the Firefly was the most efficient aircraft flown by the Fleet Air Arm. Serviceability was excellent, and its record during WW2, the Korean War and Malayan anti-terrorist campaigns, was exemplary. Powered by Rolls Royce Griffon engines giving a top speed of 370 knots the Firefly had a wing span of 41ft and was 37ft in length. Fireflies were exported to Australian, Dutch and Canadian naval air squadrons. Its roles included fighter/bomber, reconnaissance, anti-submarine and training.

Hawker Sea Fury: 801 and 804 Naval Air Squadrons were equipped with Sea Fury FB 11 whilst aboard *Glory*. It was a superb single seat aircraft which performed exceptionally well during the Korean War. The Sea Fury had a huge five bladed propeller and was powered by a Bristol Centaurus 18 engine, giving a maximum speed of 400 knots. The wing span was 38ft 4¾in and length 34ft 8in. With four 20mm cannon mounted in the wings, Sea Furies could be adapted to carry two 1,000lb bombs, 500lb bombs, 12 60lb rocket projectiles, two 90 or 45 gallon drop tanks, and an assortment of mines, depth charges, flares or markers.

Supermarine Sea Otter: 'Hands to seaplane stations' was a familiar pipe aboard HMS *Glory* during 1950. The Sea Otter, the ship's amphibious biplane, was mail plane, messenger and general run-about. When in harbour, it would be lowered into the water by crane and after a great deal of noise and a tremendously long take off run, would eventually lurch into the Mediterranean sky. Powered by a Bristol Mercury XXX 855 hp engine mounted on its top wing, the Sea Otter was 39ft 5in in length, had a wing span of 46ft, and had a maximum speed of 240 knots. It had a crew of pilot and observer and could carry two passengers. For armament two Vickers machine guns could be mounted, one amidships and one in the bows.

Supermarine Walrus: Designed by R.J. Mitchell, this amphibian first appeared in 1933 under the name Seagull V. It was adopted by the Admiralty in 1935 as the standard ABR (amphibian-boat-reconnaissance) and renamed Walrus. By the time the Second World War began, in addition to all aircraft carriers, most of the Royal Navy battleships, battle-cruisers and cruisers were also equipped with a catapult-launched Walrus. Affectionately known as the 'Shagbat', this aircraft performed a most valuable service not only in a fleet communications role but for the rescue of many ditched aircrew, often in very dangerous situations. Powered by a 775hp Bristol Pegasus VI radial engine, it had a top speed of 135mph, and a range of 600 miles. It had a wing span of 45ft 10in (13.97m) and overall length of 37ft 7in (11.45m).

Sikorsky S51 (Helicopter): The loan of a Sikorsky S51 from the United States Navy to HMS *Glory* was much appreciated by those aircrew who had the misfortune to ditch in the sea or were forced to land in hostile country during the Korean War. Later known as the Dragonfly, they served on every aircraft carrier in commission during the 1950s carrying out duties of planeguard, or search and rescue. Powered by a 550hp Alvis Leonides 50 engine, the Dragonfly had a maximum speed of 90 knots and carried a crew of two (pilot and aircrewman).

Aircraft Carrier identification: In April 1945 the Royal Navy introduced deck identification letters for its aircraft carriers. HMS *Glory* was allocated the letter 'L', and this was painted in white on the starboard side of the flight deck for'd and port side of the flight deck aft. The identification letter was later changed to 'R' and remained for the duration of HMS *Glory*'s career.

Aircraft identification: Royal Navy aircraft were also issued with identification codes which consisted of a letter and number. On joining the British Pacific Fleet, *Glory*'s aircraft were allocated the letter 'Y' which was painted on the tail fin. In addition the aircraft also carried a three digit number on each side of the fuselage for individual identification. Each aircraft carrier in the fleet used the same basic combination of codes: those in the 111 to 169 range being reserved for single seat aircraft, 270 to 298 being used for two seater aircraft, and three seaters 370 and 398. For example: from August 1945 until the early part of 1946 the code on a Corsair of 1831 Squadron would have been the letter 'Y' on the fin, and a number between 111 and 123 on the fuselage (see photograph of *Glory* entering Melbourne Harbour January 1946). In 1951 a Firefly of 812 Squadron could be identified by the letter 'R' on the fin and a number between 201 and 211.

COLOSSUS CLASS AIRCRAFT CARRIERS

HMS Colossus: Built by Vickers Armstrong (Tyne), launched on Thursday 30 September 1943. On loan to Franch Navy August 1946, and named *Arromanches*. Purchased by the French 1951. Put on disposal list 1974. Broken up in Toulon 1978.

HMS Glory: Built by Harland and Wolff (Belfast) and launched Saturday 27 November 1943. Placed in reserve 1956. Arrived Inverkeithing on Wednesday 23 August 1961 for breaking-up.

HMS Ocean: Built by Alex Stephen & Sons Ltd. (Govan), launched Saturday 8 July 1944. Placed in reserve 1958. Arrived Faslane Sunday 6 May 1962 for breaking-up.

HMS Theseus: Built by Fairfield. Launched Thursday 6 July 1944. Placed in reserve 1956. Arrived Inverkeithing Tuesday 29 May 1962 for breaking-up.

HMS Triumph: Built by Hawthorn Leslie, and launched Monday 2 October 1944. Converted to maintenance ship 1958–65. In reserve 1972, for disposal 1980. Left Chatham to be broken up in Spain, Wednesday 9 December 1981.

HMS Vengeance: Built by Swan Hunter (Wallsend). Launched Wednesday 23 February 1944. Lent to Royal Australian Navy 1953–55. Sold to Brazil 14 December 1956, underwent modernisation in Rotterdam 1957–60. Renamed *Minas Gerais* and joined Brazilian Navy 13 January 1961. In 1976–81, underwent extensive refit to give another ten years service.

HMS Venerable: Built by Cammell Laird and launched Thursday 30 December 1943. Sold to the Netherlands Thursday 1 April 1948, renamed *Karel Doorman*. Sold to Argentina Tuesday 15 October 1968 and renamed *25 de Mayo*.

HMS Warrior: Built by Harland & Wolff (Belfast), and launched Saturday 20 May 1944. Lent to Royal Canadian Navy 1946–48, and then in 1958 to Argentina and renamed *Independencia*.

HMS Perseus: Built at Vickers Armstrong (Tyne) and launched Sunday 26 March 1944. Completed as Aircraft Repair Ship. Arrived at Port Glasgow for scrapping on Tuesday 6 May 1958.

HMS Pioneer: Built at Vickers Armstrong (Barrow) and launched Saturday 20 May 1944. Scrapped September 1954.

United States Navy

AIRCRAFT CARRIERS WHICH SERVED IN KOREAN WATERS

USS Antietam (CVA 36)	USS Valley Forge (CVA 35)	USS Bairoko (CVE)
USS Bataan (CVL)	USS Badoeng Strait (CVE 116)	USS Boxer (CVA 21)
USS Essex (CVA 9)	USS Bon Homme Richard (CVA 31)	USS Leyte (CVA 32)
USS Lake Champlain (CVA 39)	USS Princeton (CVA 37)	USS Sicily (CVE)
USS Oriskany (CVA 34)	USS Philippine Sea (CVA 47)	USS Kearsarge (CVA 33)

United States aircraft carriers were classified as either ATTACK CARRIERS (CVAs), ESCORT CARRIERS (CVEs), or SMALL CARRIERS (CVLs).

UNITED STATES NAVY CARRIER AIRCRAFT

(AD) Douglas 'SKYRAIDER' Divebomber
(F9F) Grumman 'PANTHER' Jet Fighter
(F2H2) McDonnell 'BANSHEE' Jet Fighter
(F4U) Vought 'CORSAIR' Fighter.

Acknowledgements

I would like to thank all those members of the GLORY ASSOCIATION and others, for sending me newspaper cuttings, photographs, journals, diaries, letters, and personal accounts of their time served in HMS *Glory*. Their help has been of great value and I hope they will enjoy the end result, particularly:

Mr Bernard Allkins (Chairman GLORY ASSOCIATION).

Mr Dennis T. Brooks, Mr Harry Burrow, Mr Philip Lister.

Mr Alan Culbard, Mr Eric Field, Mr George Dickie.

Mr Bryan Green, Mr Dennis Hammond, Mr Anthony Hammond.

Mr Bertram Hubbard, Mr Bernard Cohen, Mrs Margaret Carlow.

Mr Jim Pickett, Mr George Reid, Mr Norman Wilcock.

Mr Derek Collie, Mr Pat Davies, Mr Tony Lloyd-Davies.

Mr Gordon Brown, Mr Bert Roach, Mr David Wharton.

Mr Malcolm Lucas (HMS *Birmingham* 1952–53), Mr David Hyam.

Mr Paul Wilson, Mrs Pamela Mountain.

Mr Richard Evans, Mrs Doris Simpson, Mr Luke Grant.

Bibliography and Sources

HMS *Glory* LOG BOOKS, Public Records Office, Kew.

837 Squadron Diary, Public Records Office, Kew.

HMS *Glory*, Report of Proceedings, Public Records Office, Kew.

Keesings Contemporary Archives.

'An Offshore Incident', by P.G.W. Morris RN (Ret.).

Articles, Facts, and Figures, from the booklet HMS GLORY, a record of 1951–53, and in particular to AVS, EFS, EHG, AW, AGFW, Sgt RMs, TS and PW.

Petty Officer W.G. Thomas, 821 Squadron (Letters to his Mother from Friday 25 September 1952 to Thursday 11 December 1952, kept at the Imperial War Museum).

Mrs Sheila Clare, Archives Assistant, Admiral Buzzard File, Churchill College, Cambridge.

Mr T. McCluskie, Harland and Wolff (Technical Services Ltd.), Queens Island, Belfast, BT3 9DU.

Mr P.R. Melton, Naval Staff Division (Historical Section), Ministry of Defence, Empress State Building, London SW6 1TR.

Sir Anthony Buzzard, 1194 Mud Creek Road, Oregon, Illinois, 601061, USA.

Mrs Blenkinsop, 46 Melrose Road, London SW18 1LY.

Mr David Brown (Air Pictorial), July 1974.

Mr Ray Sturtivant, The Squadrons of the Fleet Air Arm, Air Britain Publications, ISBN 0 85130 120 7.

Linda Haston (Librarian), Western Evening Herald.

William Macauley (Librarian), Belfast Telegraph.

Brenda Woods (Librarian), The Scotsman, Edinburgh, Scotland.

The Inverness Courier, Bank Lane, Inverness.

Shelley Fralic, Deputy Managing Director, The Vancouver Sun, Granville Street, Vancouver, Canada.

Belfast Newsletter, 1 January 1946.

The Adelaide Advertiser, 121 King William Street, Adelaide, South Australia 5001.

Ulster Folk and Transport Museum, Holywood, N Ireland.

Honor Morris, Programme Researcher, BBC Radio Kent.

Edwin G. Bowman (Librarian), The Times of Malta.

Mr Simon Robbins, Department of Documents, Imperial War Museum.

Mr Guy Robbins, Old Brass Foundry, National Maritime Museum.

Mr Charles Causley, Collected Poems 1951–75 (Macmillan).

The Sea War in Korea, Malcolm W. Cagle and Frank A. Manson, United States Naval Institute, Annapolis, USA.

With the Carriers in Korea (pages 436, 438, 440, 444, 445), John R.P. Lansdown (A Square One Publication).

British Carrier Aviation, Norman Friedman.

The Rise and Fall of the Aircraft Carrier (Bernard Ireland).

Fleet Air Arm History, Lieut.-Cmdr. J. Waterman RD, RNR (Rtd).

Fleet Air Arm (The Admiralty Account of Naval Air Operations in World War Two).

The Navy and the Y Scheme (HMSO) (Code 20.129).

Men Dressed as Seamen, S. Gorley Putt (Christophers).

Royal Naval Air Squadron Diaries, 801, 804, 812, 821, Mr L.F. Lovell and Mr M.D. Richardson (Research Officer), The Fleet Air Arm Museum, RNAS, Yeovilton, Somerset.

Ashley Cunningham-Booth, Life President British Korean Veterans Association, Hon President HMS NEWFOUNDLAND ASSOCIATION (1942 Commission).

Kate Lyall Grant, David Higham Associates, Golden Square, London.

Carrier, Printed and Published by Portsmouth and Sunderland Newspapers plc, at the News Centre, Military Road, Hilsea, Portsmouth.

The Morning Calm, Journal of British Korean Veterans Association, Number 24, April 1991.

Index

210

211

212